Advances in
HYDROSCIENCE

VOLUME 6

Contributors to Volume 6

JACOB BEAR
RICHARD L. EDDY
R. A. HORNE
ARCHIE M. KAHAN
JAN VAN SCHILFGAARDE

Advances in HYDROSCIENCE

Edited by

VEN TE CHOW

UNIVERSITY OF ILLINOIS
URBANA, ILLINOIS

Volume 6 - 1970

ACADEMIC PRESS New York and London

ACADEMIC PRESS, INC.
111 Fifth Avenue, New York, New York 10003

United Kingdom Edition published by
ACADEMIC PRESS, INC. (LONDON) LTD.
Berkeley Square House, London W1X 6BA

LIBRARY OF CONGRESS CATALOG CARD NUMBER: 64-17306

PRINTED IN THE UNITED STATES OF AMERICA

Contents

TWO-LIQUID FLOWS IN POROUS MEDIA

JACOB BEAR

List of Contributors

JACOB BEAR, Technion-Israel Institute of Technology, Haifa, Israel

RICHARD L. EDDY, Office of Atmospheric Water Resources, U.S. Bureau of Reclamation, U.S. Department of the Interior, Denver, Colorado

R. A. HORNE*, Arthur D. Little, Inc., Cambridge, Massachusetts

ARCHIE M. KAHAN, Office of Atmospheric Water Resources, U.S. Bureau of Reclamation, U.S. Department of the Interior, Denver, Colorado

JAN VAN SCHILFGAARDE, Soil and Water Conservation Research Division, Agricultural Research Service, U.S. Department of Agriculture, Beltsville, Maryland

*Present address: Woods Hole Oceanographic Institution, Woods Hole, Massachusetts

Foreword

The articles selected for this volume are related to water occurring in air, on land and in the ocean. Within the broad scope of hydroscience, the topics so presented thus encompass all three parts of the total water system on earth; namely, atmosphere, hydrosphere, and lithosphere. It is hoped that this wide coverage of material would benefit the graduate scholar embarking upon his research career, the scientist or engineer interested in changing his research emphasis, and the research director in keeping up with current scientific and technological advances. Reviews found in this volume, like those in previous volumes, are written by individuals actively engaged in research and as such should reflect the authors' perspective and judgment as to what is important both in recent accomplishments and in proposed problems.

Air and water are by far the most common elements in our environment on earth. As water is always in presence in air and it is constantly carried over land areas by the latter, atmosphere is obviously a vast potential source of water supply. Various ways of "rain-making" to extract this water have been attempted, but past results can hardly be considered as satisfactory mainly because of insufficiency in basic scientific research and development on this problem of a very difficult nature that has been little understood. In recent years, a more basic study of this difficult problem has been made possible due to new scientific and technological developments, including improving cloud-seeding techniques, new knowledge in cloud physics, great progress in mathematical models by high-speed computers, significant advances in modern statistical procedures for evaluating the theories and experiments, and the emergence of new capabilities such as the use of artificial satellites and remote sensing to observe and measure atmospheric physical parameters. The Weather Modification Act passed by the United States Senate in 1966 has further stimulated research activities in this area. The article by Kahan and Eddy should be appreciated as it provides the reader an authoritative review of new developments in atmospheric water resources and also an encouraging outlook and needed research possibilities for the future.

The article by van Schilfgaarde deals with the theoretical aspects of drainage of water on land which is an important subject, as pointed out by its author, that relates particularly to agriculture for food production. Traditionally, engineering design of drainage systems is largly based on empirical and semiempirical criteria and assumptions because of the inavailability of sound theories. This article expertly summarizes the adequate drainage theory that has been developed in recent years, which can serve as a scientific basis for practical drainage designs.

Also related to water on land is the article by Bear, which presents the basic concepts, principles, and methods of analysis for two liquid flows in a porous medium domain. The discussions so presented must be of interest to ground water hydrologists and geologists and also to petroleum engineers who deal with similar problems in reservoir engineering for oil explorations.

With recent interests in oceanology and ocean engineering, a timely article is prepared by Horne on sea water. This article gives comprehensive scientific information on the constituents, the physical–chemical properties, and the structure of sea water that covers more than two-thirds of the earth surface. Such information is undoubtedly of vital importance to scientists and engineers who venture to explore the inner space of our earth.

Contents of Previous Volumes

VOLUME 3

VOLUME 4

VOLUME 5

Advances in

HYDROSCIENCE

VOLUME 6

NEW DEVELOPMENTS IN ATMOSPHERIC WATER RESOURCES

ARCHIE M. KAHAN and RICHARD L. EDDY

Office of Atmospheric Water Resources
U.S. Bureau of Reclamation
U.S. Department of the Interior
Denver, Colorado

I. Introduction

From one point of view any development in the area of atmospheric water resources is a new development. Although man's efforts to increase his water supply by operating in the atmosphere are not of recent origin, the concept that the atmosphere contains a water resource that can be managed in much the same way as other resources is a comparatively recent idea.

The atmosphere contains a vast amount of water, most of which is in the form of vapor. There are about 1.5×10^{16} gallons (4.7×10^{10} acre-feet) of moisture passing across the United States during an average year, about one-tenth of which will fall out as precipitation. While these average values obscure the facts that precipitation is not evenly distributed in either time or space, and that some of the precipitation falls in areas where it has not been used by man, they do suggest the tremendous potential of atmospheric water resources.

The effort to learn how to manage the atmospheric water resources is coming none too soon. The 1968 report of the U.S. Water Resources Council [*1*] identified regions with present and predicted water problems. Take the Rio Grande Region as an example. With an average natural runoff of about 5 billion gallons per day, it will have for the year 1980, an estimated withdrawal of over 8 billion gallons per day. This Region currently suffers not only from water supply deficiencies and groundwater storage depletion but also from poor quality because of mineral pollution, heavy sediment loads in many tributaries, excessive nonbeneficial losses of water, and frequent damaging floods. Three other regions—the Upper and Lower Colorado and the Great Basin—are already experiencing similarly severe water supply problems in some parts. Four more water resource regions—the Texas-Gulf, the Arkansas-White-Red, the Missouri, and the Souris-Red-Rainy—already have major water supply problems in some areas.

Water requirements will increase as the population increases. By the year 2000 the population of the Southwest United States is expected to exceed 31 million. These water supply problems (including quality as well as quantity) must be solved if this area is to support the anticipated rates and patterns of economic growth.

The solution to water problems will not come solely from increased quantities. Many parts of the world have abundant quantity, but pollution has seriously reduced the usefulness of the supply. In a few places, such as Washington and Oregon, redistribution of water, rather than increased quantity, would greatly ease the water resource situation. The ability to manage precipitation, if only to a small degree, could increase the amount of water stored in reservoirs, could increase the flow in rivers and help dilute pollutants, and could redistribute moisture by retarding precipitation processes on one side of a mountain and stimulate them on the other side.

Efforts to manage precipitation are certainly not new. First attempts were probably through appeals to deity. The first scientific discourse was J. Boecler's *Dissertatio Physica Pluvia*, published in 1710. In the United States, James P. Espy deserves recognition. In 1839, he published a paper suggesting that droughts could be broken by burning large areas of salt marshes, prairies, or timber. He recognized the importance of vertical motion in the precipitation process and reasoned that the heat produced would generate upward currents of air and result in clouds.

During the last half of the 19th century, a substantial number of scientific rainmaking ideas were submitted. One theory involved modifying atmospheric electricity, another cooling the air to cause condensation, and still another suggested that the concussion from aerial explosions would cause condensation. In 1895, Alexander McAdie suggested that supercooled clouds (clouds at temperatures lower than 0°C) could be triggered into raining by putting ice crystals into them. The importance of this concept apparently escaped the scientific minds of that day.

Shortly after the turn of the century, Charles Hatfield claimed that his fume-producing chemicals would make rain. He was hired by a number of organizations in California to solve their water problems. One such organization was the City of San Diego. His rainmaking efforts were followed by one of the heaviest rains on record, so heavy, in fact, that serious flooding resulted. His reward was the threat of a lawsuit. The constitutents of his rainmaking formula were never made public. He and other rainmakers were criticized, perhaps too harshly, by some members of the scientific community.

It was not until the 1940s that methods based on a knowledge of the physical processes of precipitation formation were used. In 1946, Vincent J. Schaefer discovered that a small piece of dry ice, when dropped into a cold chamber filled with supercooled water droplets, caused the formation of millions of tiny ice crystals. (In Holland in 1930, A. W. Veraart dropped dry ice, among other things, into supercooled clouds, but apparently did not notice any ice crystals.) The first field experiment was conducted in November 1946, with definite changes in cloud characteristics observed. Also in November 1946, Bernard Vonnegut discovered that minute crystals of silver iodide acted as ice-forming nuclei at temperatures lower than -5°C. The first large-scale trials using silver iodide took place in June 1950.

During the late 1940s and early 1950s, dozens of rainmakers appeared. Difficulties in evaluating results led to some skepticism. In 1953, Congress established the Advisory Committee on Weather Control to study and evaluate public and private experiments in weather modification. The Committee published its findings in 1957 [*2*]. The Committee concluded that a statistically significant 10% to 15% increase in precipitation resulted from seeding winter storms in mountainous areas of the Western United States, but no increases were found in nonmountainous areas.

The activities of the Advisory Committee and the general cautiousness of the nation with regard to operational cloud seeding led to a period of retrenchment. Basic research, aimed at understanding more fully the fundamental concepts, increased. The field projects that were carried out were evaluated more thoroughly, and a much greater effort was directed toward getting scientific data.

In 1963, the National Academy of Sciences appointed a Panel on Weather and Climate Modification, charged with a "deliberate and thoughtful review of the present status and activities in this field, and of its potential and limitations for the future." The Panel's final report, published in 1966 [*3*], concluded that "precipitation from some types of cloud and storm systems can be modestly increased or redistributed by seeding techniques." The increases found by the Panel were on the order of 10%. Both the Advisory Committee on Weather Control and the National Academy of Sciences recommended increased research efforts in weather modification.

Several agencies of the Federal Government have important programs in precipitation management research. The Department of Commerce, through its Environmental Science Services Administration (ESSA), is studying winter storms over the Great Lakes. The goal is to spread lighter snowfall over a larger inland area instead of the large accumulation of snow on the lee shores of the lakes. ESSA and the U.S. Navy are looking at modification of tropical cumulus clouds. *Project Stormfury*, another joint ESSA–Navy experiment, is not specifically a precipitation management program, but does involve seeding portions of hurricanes. Other ESSA programs related to precipitation management include computer modeling, inadvertent modification studies, and a benchmark observation program for long-term changes in the atmosphere.

The National Science Foundation (NSF) supports most of the basic research in weather modification. A number of cloud physics, cloud dynamics, and instrumentation development projects are contributing substantially to the technology of precipitation management. NSF has also supported major field experiments in cloud seeding. In addition, NSF supports research in fields affected by weather modification, such as ecology.

The Department of the Interior, through its Bureau of Reclamation, concentrates on applied precipitation management research. *Project Skywater*, as the program is called, is charged with the responsibility to develop precipitation management into an effective method for improving the nation's water resources. Almost all of the research is conducted through contracts with universities, private research groups, and state agencies. Other Federal agencies also participate in the Bureau of Reclamation program.

Other Federal Government agencies have programs closely related to precipitation management. The Forest Service of the Department of Agriculture has a program in lightning suppression. The Department of Defense, the Federal Aviation Administration, and the National Aeronautics and Space Administration are studying fog dispersal. ESSA and NSF are investigating hail suppression and severe storm abatement. Total Federal funding for all weather modification research has been on the order of $10 million a year, with precipitation management accounting for less than half that total [*4*].

A relatively small, but important, contribution to precipitation management research is being made by several private industrial groups. Power companies and municipal water districts, especially on the West Coast, have supported cloud seeding operations for many years. Much of the present precipitation management technology comes from projects like these.

II. Statistically Designed Cloud-Seeding Experiments

A. Early Experiments in the United States

1. *The Santa Catalina Mountains, Arizona Experiment*

a. DESIGN OF THE EXPERIMENT. This experiment [*5, 6*] used an objective forecast technique based upon the precipitable water between the ground and 400 mb as observed by a 5:00 a.m. radiosonde release. If the precipitable water exceeded 1.10 inches, the day was considered suitable. On the first such day, an envelope was opened which stated "seed today—do not seed tomorrow" or the reverse. The ordering of the envelopes followed a randomization scheme. One suitable day was allowed between members of a pair. Seeding was done by an airborne silver iodide generator. Seeding took place for a 2- to 4-hour period along a line perpendicular to the wind at flight altitude and about 30 minutes upwind from the target area. This program, Program I, was conducted during the summers of 1957 through 1960. In Program II, the summers of 1961 through 1964, the essential design of the experiment did not change, except that the aircraft flew 1000–2000 ft below the cloud base and a 10%, rather than a 2%, silver iodide solution was used.

b. RESULTS OF THE EXPERIMENT. No increase in rainfall at the ground was indicated. However, analysis of radar data indicated that the precipitation process was initiated earlier in those clouds which were seeded, even though no significant increase in rainfall was detected.

2. *The Project Whitetop, Missouri Experiment*

a. DESIGN OF THE EXPERIMENT. This study [*7, 8*] was directed toward summer, nonorographic, convective clouds. Silver iodide nuclei were released at cloud-base levels from airborne generators on a random sample of operational days. Operational days were defined in terms of a previously selected set of meteorological parameters. Seeding was done along a 30-mile-long line perpendicular to the wind along the upwind side of a 120-mile circular area.

b. RESULTS OF THE EXPERIMENT. Based upon an analysis of radar data, the following suggestions were made: (1) Cloud-base seeding has a direct effect upon the larger-than-average clouds in the immediate vicinity of the seeding plume. (2) A result of the increased growth of the clouds in the seeded area is the formation of a stationary-wave-like perturbation which spreads outward and downstream with some sort of damped harmonic motion with a wavelength of 30–50 miles. (3) This wave creates a region unfavorable for cloud growth roughly 40 to 80 miles downwind and with a favorable region beyond about 80 miles.

3. *The Santa Barbara, California Experiment*

a. DESIGN OF THE EXPERIMENT. The experiment [*9*] was designed as a randomized experiment in which roughly half of all seeding opportunities in Santa Barbara County would be treated and the remainder would be left untreated to serve as a control. In this experiment, one organization conducted the program, another group collected the data, and a third performed the evaluation of the results of the program. The "unit of observation" was a 12-hour period. Some time prior to the beginning of each unit of observation, a forecast was made as to whether or not the unit would be a suitable seeding opportunity. If suitable, a random decision of either seed or no-seed was made and carried out.

This plan was carried out for 1957. In 1958, the adjacent Ventura County joined the project, but due to drought conditions, Ventura County was seeded at every opportunity. In 1959, both counties operated under the conditions of the original plan, but the year was very dry and only nine seeding opportunities occurred.

b. RESULTS OF THE EXPERIMENT. Due to the fact that the original plan was not followed for the program test period, no statistically reliable results are apparent.

B. RECENT EXPERIMENTS IN THE UNITED STATES

1. *The Climax, Colorado Experiment*

a. DESIGN OF THE EXPERIMENT. The basic features of the experimental design [*10*, *11*] are as follows:

(1) Randomization is used to obtain the seeded and unseeded samples.

(2) The time unit is 24 hours. This is a compromise which minimized variations in the physical parameters during an event, but is still long enough to decrease the "noise" level to acceptable values for establishing correlations with upwind controls.

(3) Observations of meteorological variables are made as intensively as feasible during all stages of the precipitation process in both seeded and nonseeded cases.

The operational procedures are as follows:

(1) Randomization is accomplished by drawing 100 paired slips from a container at the start of each season. A chronological ordering of decisions is prepared.

(2) The suitability of each day is determined by the Denver Weather Bureau Office. The criterion of a suitable day is that Leadville, Colorado, is expected to have at least 0.01 inch of precipitation in the 24-hour period. The forecast is made 6 to 8 hours prior to the start of each experimental period. The forecasters have no access to the seeding decisions or specifically how they will be used.

(3) The "yes" or "no" for suitable experimental conditions is relayed from the Weather Bureau to Colorado State University where the forecast is processed and action is taken to initiate seeding operations if the next consecutive listing of a random decision is for an operational day.

(4) Generator operators at the sites closest to the target area turn on the generators 30 minutes prior to the start of the experimental period and terminate operations 30 minutes prior to the end of the period. The same procedure is used for the more distant sites but a 1-hour lead time is used. The specific generators used are determined by the Weather Bureau wind forecast. A CSU-modified Skyfire, needle-type ground generator is used for seeding. The generators have been extensively calibrated in the CSU cloud chamber to establish the temperature activation characteristics of the nuclei produced. Consuming about 20 gm of silver iodide per hour, the units produce about 10^{14} particles/gm of AgI effective at $-12°C$ and 4×10^{15} particles/gm effective at $-20°C$.

(5) Some 70 snow boards, located at about 1-mile intervals over three mountain passes in and near the target area, are read daily. At each snow board, the snow depth is measured, the weight of a snow core is determined, a snow sample for water analysis is secured, and drifting or melting, if any, is noted. The snow boards are read in a systematic order each day, and the observers have no knowledge as to seeding operations.

(6) As many physical observations as possible are made on both seeded and unseeded days. Ice nuclei observations have been made on almost all days. A detailed weather observation is taken at the time of each ice nuclei observation. Various other observations are made whenever possible. These include ice crystal replicas, cloud and radar photography, and atmospheric electricity. In 1965, circumstances forced a change in the location of certain generator sites. Since the basic philosophy remained the same, the experiment was continued, but was named *Climax 2*, to differentiate the relocated generators from the original sites used in *Climax 1*, which was conducted from 1959 through 1964.

b. SUMMARY OF RESULTS. Data for the period February 1960 to April 1968 were recently analyzed [*11*]. Table I shows the percentage change in precipitation between seed and nonseed days for Climax 1 and Climax 2 stratified by 500-mb temperature and adjusted for control station precipitation. All values shown are statistically significant. These results clearly indicate the strong temperature dependency for seeding effectiveness.

TABLE I

PERCENTAGE CHANGE IN PRECIPITATION, CLIMAX EXPERIMENT ADJUSTED FOR CONTROL AREA PRECIPITATION

500-mb Temperature (°C)	Climax 1	Climax 2
−15°	+96%	+171%
−20°	+26%	− 39%
−25°	−22%	− 1%
−30°	−29%	− 70%

2. *The Lake Almanor, California Experiment*

a. DESIGN OF THE EXPERIMENT. Six silver iodide generators were located at elevations between 6000 and 7500 ft on ridges or near mountain tops on the southeast, south, and west sides of the Lake Almanor watershed [*12*]. The generators burned about 25 gm of silver iodide per hour and produced approximately 10^{15} nuclei/gm, effective at −15°C. The units were operated by remote control from a powerhouse about 12 miles south of the lake.

During the period January 28, 1963, through May 14, 1963, orders to seed were issued by the Pacific Gas and Electric Company weather forecaster in San Francisco. Every order was an order to seed in that either the east group of two burners or the west group of three or four burners was operated. The group specified was a random decision made before the season started, and of which the forecaster had no prior knowledge.

Wind, temperature, and precipitation were the principal measurements made on the watershed. Wind was measured at the 5000- and 7500-ft levels on the watershed. Additional wind information came from radar photography and from upper air weather charts. Temperature was measured at each generator site and at a location within the watershed. Free air temperature soundings were made at times during seeding. Precipitation was measured by 51 heated weighing gages spaced at approximately 2-mile intervals on the watershed.

Precipitation data were stratified by wind direction and temperature. Wind from 160° through 220° was classified as south wind; from 230° through 300°, west wind. If the −5°C level was below 7500 ft, the situation was termed cold. If the −5°C level was higher than 7500 ft, it was classified as a warm situation.

b. RESULTS OF THE EXPERIMENT. Three different analyses—ratio, regression, and covariance—gave an approximate increase in precipitation of 60% on seeded days for the cold west condition. When the warm and cold west conditions were combined, a 40% increase in precipitation was noticed.

No effects comparable to the westerly situations were noted for the southerly wind conditions. In particular, no positive changes were observed that could be identified with significantly different seeded and unseeded regression lines. It is thought that the thermally stable air often associated with southerly flow may have been a factor in these cases. Stable air might prevent the nuclei from reaching the clouds at the levels and in the amounts required for effective nucleation.

3. *The Rapid City, South Dakota Experiment*

a. DESIGN OF THE EXPERIMENT. A randomized crossover design was employed in this experiment [*13*]. In this design, a pair of target areas is selected and one or the other is seeded on a given test case according to a random decision. Since two distinct patterns of shower activity, dependent upon northwest or southwest flow aloft, had been observed, two pairs of target areas were selected. This procedure minimized the possibility of contaminating the unseeded target area, since one pair was used on days with northwesterly flow and the other pair used with southwesterly flow.

To determine seed and nonseed situations and to permit stratification of the data, rules were devised which placed the weather conditions for a day into one of four types. These types were superimposed upon the two types of wind flow so that eight combinations were possible.

b. RESULTS OF THE EXPERIMENT. Analysis of covariance, rank tests, and rainfall ratios of 98 test days occurring during the summer seasons of 1966, 1967, 1968 led to the following conclusions:

(1) On days with isolated showers, rainfall is heavier in the seeded target area than in the unseeded target area.

(2) On days with widespread convective activity and southwesterly winds aloft, rainfall is lighter in the seeded target area, except for the region 10 to 20 miles east of the Black Hills.

(3) On days with widespread convective activity and northwesterly winds aloft, rainfall is lighter in the seeded than in the unseeded target areas.

4. *The Elko, Nevada Experiment*

a. DESIGN OF THE EXPERIMENT. The experiment was started in 1961 [*14*, *15*]. It was designed as a wintertime random crossover seeding experiment with a target area in the Ruby Mountains and one in the Jarbridge Mountains of Elko County. By seeding one or the other of the targets on a random basis, it was estimated that 75 trials would be sufficient to determine the effectiveness of ground-based silver iodide seeding in the area.

b. RESULTS OF THE EXPERIMENT. After 58 trials were completed, an interim statistical analysis indicated a slight, but not statistically significant, increase in precipitation in the target area. The originally planned 75 trials would be insufficient to attain significant levels of confidence, and a prohibitive number of trials would be required. Analysis of the silver content of 59 snow samples from the target areas indicated that during the 1967–68 winter season, the silver iodide released from ground generators did not fall within the target area. Thus, the apparently low targeting efficiency of the project precluded any statement regarding the efficiency of the seeding effort.

C. EXPERIMENTS ABROAD

1. *Australia*

During the period 1955–1963, three separate experiments were conducted in Australia [*16*]. These experiments were titled New England, Snowy Mountains, and Warragamba. All employed airborne seeding in one of two areas according to a random decision. The consensus of the three experiments was that cloud seeding increased precipitation as measured at the ground. At the present time, an experiment is in progress in Tasmania to investigate the long-term carryover effects of silver iodide seeding.

Starting in 1965, Australia has undertaken practical applications of cloud seeding. These efforts are directed toward increasing precipitation for improved wheat yields and to decrease forest fire hazards.

2. *Israel*

A randomized cloud-seeding experiment has been in operation in Israel since 1961 [*17*, *18*]. Seeding is done by airborne silver iodide generators. On each day, one of two experimental areas is randomly designated and, if suitable clouds are found, seeding is carried out. The experiment is conducted during the rainy season, and is evaluated by a crossover comparison technique. Based on data obtained for five and one-half seasons, there has been a seeding-induced increase of about 19% in rainfall.

3. *Russia*

From information appearing in the technical press, it seems that most of the efforts toward precipitation management in Russia are directed toward hail suppression. Furthermore, these efforts have apparently been quite successful. Convective clouds capable of becoming hail producers are monitored by radar. When such clouds have grown to be potential hailstorms, the nucleating material is introduced into the cloud at the point determined by radar by specially designed artillery projectiles or rockets. Both silver iodide and lead diiodide have been used as the nucleant. Some attention has been given to the clearing of cold fog from airfields by silver iodide seeding from aircraft. Only minor efforts have been given to cloud seeding for precipitation increase.

4. *India*

An important warm cloud seeding experiment in India involved the release of salt from the ground during the three monsoon months—July, August, and September [*19*]. Three target areas were set up and seeding days were selected on the basis of a randomization procedure. Comparisons were made of the ratio of target area rain to control area rain for two groups of seedable days—one group seeded and the other not seeded. The data were subjected to three statistical tests, all of which indicated that the precipitation increases in two of the test areas on seeded days were statistically significant. When data for all three areas were combined, the results were highly significant. Precipitation was increased by at least 21% in this experiment.

III. Long-Term Commercial Projects Aimed at Increasing Precipitation

A. The Kings River Project

The purpose of this project [*20*] is to increase the flow of California's Kings River by cloud seeding over that portion of the drainage basin above Pine Flat Dam. Both airborne and ground-based seeding have been employed. Ground-based generators dispense silver iodide. The aerial seeding has released both silver iodide and dry ice. Since the program began in 1954, a total of about 309,000 gm of silver iodide were used and nearly 8500 gm of dry ice dispensed through the 1965–66 season.

Evaluation is based on a formula which relates the unregulated flow of the Kings River to the combined flow of the Merced and Kern Rivers. The correlation coefficient for this multiple regression is 0.985. Using the formula, there has been an apparent average annual increase in the flow of the Kings River of 76,760 acre-feet/year, at the 99% confidence level.

B. The San Joaquin River Project

This project, probably the longest continuing cloud-seeding project, has been in operation since January 1951 [*21*]. The purpose of the program is to increase runoff in the upper reaches of the San Joaquin River by seeding clouds over the Big Creek watershed. Since the inception of the project, there have been several changes in the number and location of silver iodide generators; however, 12 have been used since 1960.

Since increased runoff is the purpose of the program, the program was evaluated through runoff analysis. Regression equations were developed for the runoff of the San Joaquin River against the Merced River for the period 1921–1950 and the runoff for the seeded years were plotted on the regression line. The results indicate that there has been an 8% increase in runoff during the period 1950–1965 as a result of seeding, with a 97% probability that the increase is not due to chance.

C. Others

There are a number of other long-term commercial projects in this country which are directed toward increasing precipitation. Space limitations preclude a discussion of these programs, but, since many of them have been underway for several years, it would seem that satisfactory results are being obtained.

IV. The Technology of Precipitation Management

Results of research and operational programs since the late 1940s have identified procedures that appear related to precipitation increases. At the same time, these results have indicated areas where real competence is insufficient.

A number of cloud-seeding techniques have been developed. Ground-based seeding, using any of a variety of silver iodide generators, is the most commonly used commercial and experimental technique, especially for winter storms in mountainous terrain. The seeding material is carried aloft by vertical motion resulting from the instability of the air or from the lift due to the mountain barrier. A fundamental problem that still exists involves diffusion of the seeding material. Proper seeding procedures require (1) that the proper number of nuclei reach the effective level in the cloud, and (2) that the effect of the seeding will be felt in the desired location on the ground. The diffusion process is a rather complex function of vertical temperature distribution and the three-dimensional wind field.

Airborne seeding, using a variation of the ground-based silver iodide generator, or using crushed dry ice, is frequently employed with summer convective storms. Research programs often use aerial seeding because the seeding material can be introduced directly into the cloud at a specified location and the aircraft can carry other instruments and equipment needed to study the effects of the seeding on various cloud factors. The primary limitation of aerial operations is the capability of the aircraft to fly in weather conditions where seeding will be effective.

Various experimental designs and statistical evaluation procedures have been used, some of which, in retrospect, were inadequate for the situations that existed. Nevertheless, the early programs did show that cloud seeding has a tremendous potential. The design of either a research or an operational program must include the following: (1) clear, precise, and complete definition of the particular problem at hand; (2) thorough review of the "tools" available (where the tools can be equipment, instrumentation, seeding techniques, diffusion equations, etc.); and (3) careful attention to matching the techniques (or tools) needed to form a balanced system to accomplish the job. The best system is not necessarily composed solely of the most highly developed techniques. A crude analogy may help to make this clear. When designing and building a sports car, the engine, transmission, suspension, and brakes are chosen so that maximum performance and handling are achieved. An automatic transmission developed for a luxury car would probably not be selected, no matter how fine it might be, because such a transmission would not fit into the sports car system. When designing a seeding program for winter orographic storms, an aircraft seeding technique would probably not be selected, even though the technique is well developed. Experience has shown that the clouds apparently most susceptible to seeding are also the clouds that present serious aircraft icing problems. In addition, the level at which seeding material should be released is frequently below the level of the highest peaks, presenting a navigational problem that a reasonable pilot would choose to avoid.

The wide variety of conditions under which clouds and storms occur, plus the different type of topography over which these clouds develop, show that several, perhaps many, procedures must be available to get the best results from every situation. The obvious goal for weather modification research, considered in total, is to find the best system for any given situation. It is unlikely, however, that the real world will ever see a truly "best" system for all conditions. A reasonable procedure, short of finding the absolute "best" way, is to put the available techniques, equipment, and instrumentation together in such a way that under the existing conditions the desired effect is maximized. In other words, optimize the available systems. What, then, constitutes a system?

A true precipitation management system, even a crude and inefficient system, will consist of four major components: (1) a component to analyze the present and expected water needs and water sources, plus the anticipated effects of precipitation management on such factors as the economy and ecology of the area in question, and arrive at a decision to employ precipitation management techniques; (2) a component to recognize a weather situation where the application of precipitation management techniques would result in the desired effect (and also, hopefully, those situations in which the result would be deleterious); (3) a component to select the proper treatment material and delivery system for the situation at hand; and (4) a component to assess the results of the treatment in terms of useful water on the ground. Each of these components will now be examined in more detail.

A. Analyzer Function

The first activity of an operational system is to determine when the application of precipitation management techniques could contribute to the resolution of a water resources problem of a particular area. The nature of the water need is clearly specified by answering the questions, why is more (or perhaps less) water needed, how much is needed, where and when is it needed, and what water quality is needed? Consideration is given to storage, transportation, and treatment facilities available or required for the area in question.

After the specific need is defined, the various potential sources of additional water are examined to find the best way to fill the need. Alternative sources, such as water mining, desalination, transbasin diversion, and precipitation management, are reviewed concurrently so that the relative potential of each source can be determined. The area climatology is studied to identify the frequency of treatable cloud systems and to provide an estimate of the water potential should cloud seeding be used. The effects of the application of precipitation management techniques on the economics, ecology, and sociology of the area are examined. Is the potential cost/benefit ratio favorable? Will additional moisture adversely affect the native plant and animal communities? Will the residents of the area object to such a program?

Another important consideration is whether or not the increased precipitation would fall where a substantial portion of it would eventually be usable (if, for example, the precipitation falls on a desert where it is quickly evaporated, the program could be considered successful scientifically because more precipitation was produced, but unsuccessful in terms of beneficial water use). There are also legal questions that must be looked at, such as ownership of the land being affected, ownership of the moisture being withdrawn, licensing and indemnification procedures, and reporting procedures.

When all the available information has been considered, a decision is made. Precipitation management techniques may be inappropriate for a variety of reasons or they may be the only techniques available. However, where serious and long-term water problems exist, such as in the Lower Colorado and Rio Grande River basins, precipitation management will almost certainly be used in addition to other methods of acquiring additional water.

B. Opportunity Recognizer Function

The second major activity of the operational precipitation management system, given that the result of the analyzer function is a decision to use precipitation management techniques, is to identify those situations that can be successfully treated by the technology available. This involves a number of meteorological factors. The roles of these factors will be more clear if some of the fundamental concepts and equations of nucleation and precipitation formation are presented. More complete information is available in books devoted to cloud physics and precipitation processes [*22–24*].

1. *Fundamental Concepts*

Nucleation (condensation, ice deposition, or freezing) can be either heterogeneous or homogeneous. Heterogeneous nucleation takes place on the surface of foreign particles or on larger surfaces of water droplets or ice crystals. Homogeneous nucleation takes place on embryos made up of relatively compact collections of a few molecules of water or ice. Nucleation, heterogeneous or homogeneous, is also categorized by the resulting phase, i.e., whether the end product of the nucleation process is a liquid droplet or an ice particle. Materials effective in converting water vapor to liquid are, in general, not the materials effective in freezing the droplets or in converting water vapor directly to ice.

Consider first the condensation process. Homogeneous nucleation of water droplets requires supersaturation of several hundred percent, but a number of naturally occurring substances can initiate condensation at only a few percent supersaturation. Some of these heterogeneous nuclei are hygroscopic, such as tiny particles of sea salt, while others are nonhygroscopic, such as solid particles from combustion processes. These substances are almost always present in the air in abundant quantities, so heterogeneous nucleation is dominant in the condensation process.

Now consider the formation of ice particles. An ice particle can form by deposition of water vapor on a suitable nucleus (an ice crystal, an ice embryo, or a foreign particle) or by the freezing of a liquid droplet. Water droplets can exist in the atmosphere at temperatures as low as about −40°C. This temperature is sometimes called the *spontaneous freezing point of water*, and at −40°C the supercooled droplets freeze even in the absence of foreign particles (homogeneous nucleation). However, a number of naturally occurring substances, such as tiny clay particles, are effective as ice-forming nuclei at temperatures much higher than −40°C. Unlike condensation nuclei, ice-forming nuclei are sometimes relatively scarce. This fact is one of the fundamental concepts of precipitation management.

c. GROWTH OF WATER DROPS BY COALESCENCE. The basic equation for the growth rate of a falling drop growing by coalescing with other drops is

$$dM/dt = \pi r^2 U_t X E_1 E_2 \qquad (4)$$

where U_t is the terminal velocity of the falling water drop; X, the mass of water in the droplets in a unit volume swept out by the falling drop (volume is $\pi r^2 U_t \times$ time); E_1, *collision efficiency*, i.e., the fraction of droplets that actually collide with the falling drop; E_2, *coalescence efficiency*, i.e., the fraction of colliding droplets that stick to the falling drops; and $E_1 E_2$ is called *collection efficiency*.

The growth rate can be expressed in terms of the radius, r, and the density of the liquid, ρ_L, instead of the mass, since $M = 4\pi r^3 \rho_L/3$.

d. GROWTH OF SNOWFLAKES BY COLLISION. The basic equation for the growth of a snowflake by collision with ice crystals is

$$dV_s/dt = A_s E N_i V_i(U_s - U_i)$$

where V_s is the volume of the falling snowflake; A_s, the effective area of the snowflake; E, collection efficiency; U_s, the terminal velocity of the snowflake; U_i, the terminal velocity of the ice crystals; and N_i, the concentration of ice crystals of volume V_i smaller than the collector snowflake.

This expression assumes the terminal velocities of both the collector snowflake and the collected ice crystals are independent of size. It also assumes the collected crystals add their volume to that of the collector without change in density.

3. *Nuclei Requirements*

The discussion of the precipitation processes stated the need for nuclei but did not indicate how many are needed to produce measurable precipitation. While no completely satisfactory answer can be given at this time to that question, one range of values frequently mentioned for ice-forming nuclei is 10^3 to 10^4 nuclei per cubic meter at the effective level in the cloud. The effective level can be defined loosely as the level at which the material is active as ice nuclei. If fewer than 10^3 nuclei are available, the precipitation process (ice crystal theory) will be inefficient or essentially inoperative. If many more than 10^4 are present most, if not all, of the supercooled droplets will be converted into tiny ice particles and few will be large enough to fall.

The number of ice nuclei in the air can be measured with any of several types of counters. Some of these instruments can be mounted in aircraft in order to obtain three-dimensional distributions. The expense of aircraft operations has thus far precluded routine sampling over large areas for extended periods of time. Eventually, however, ground measurements, plus airborne counts, will become routine because of the importance of the number of natural nuclei not only to weather modification but also to weather forecasting in general.

In many locations, man's activities are creating sources of condensation and ice nuclei. Certain industrial operations release to the atmosphere materials that are effective condensation and ice nuclei. Under some weather conditions the concentrations of these materials are extremely high and can inadvertently modify the weather over a fairly large area. Lead from the exhaust of automobiles, if combined with free iodine that may be present in the air, can form PbI_2, an effective ice-nucleating material [*26*, *27*]. Heavy traffic areas are often also industrial areas where free iodine is likely to be present. Schaefer [*28*] presents evidence that inadvertent modification is significant in some portions of the Eastern United States already. These intermittent and nonrandom sources of nuclei need to be considered in establishing the treatability of a given storm situation.

4. *Cloud Dynamics and Thermodynamics*

The three-dimensional wind field and the vertical temperature and moisture distributions determine in large measure the treatability of a given situation. If cloud seeding material is released from the ground, the vertical component of the wind field transports the material to the effective level in the cloud. The vertical motion is determined by the stability of the air (in turn determined by the vertical temperature structure), by air being forced to rise over some barrier (such as a mountain or a front), or by both. Around sunrise the surface of the earth is frequently colder than the air several hundred feet above it. This inversion condition, under most circumstances, prevents seeding material released from the ground from reaching the cloud.

The vertical moisture distribution is important because it shows how much moisture is present and at what altitudes it can be found. There is mounting evidence that the temperature at cloud top level is closely related to the effectiveness of seeding (using silver iodide). If the cloud top, determined from the moisture distribution, is too cold, seeding will be ineffective, or may actually decrease precipitation. Similarly, if the cloud is too warm or if the cloud is quite shallow, seeding will not be effective.

When vapor condenses or water droplets freeze, the released latent heat is added to the air in the vicinity of the particles. If the amount of heat is large enough in a relatively small volume in a cloud, the temperature increase will give rise to buoyant forces and result in significant vertical motion, sufficient to stimulate further growth of the cloud and process more water through the system. If the temperature and moisture distributions are known, the buoyancy, additional cloud growth, and additional precipitation can be calculated using equations such as those presented in Section IV, B, 2. Computer models now permit these and other computations to be made with a minimum of lag between observation and application.

5. *Instrumentation Requirements*

The large number of factors that need to be considered in identifying treatable situations requires many specialized instruments and pieces of equipment.

FIG. 1. NCAR instrumented aircraft typical of the type used in weather modification research and operations. Note the probes extending through the aircraft nose.

Both direct-sensing instrumentation, such as thermometers and nuclei counters, and remote-sensing instrumentation, such as radar and radiometers, are desirable. The specific instruments used are too numerous to mention individually, but the instrumentation requirements can be divided into four categories:

(1) Instruments to measure microphysical factors.
(2) Instruments to measure fluid and thermodynamical factors.
(3) Instruments to measure air mass factors.
(4) Instruments and equipment to transmit, process, and display the data,

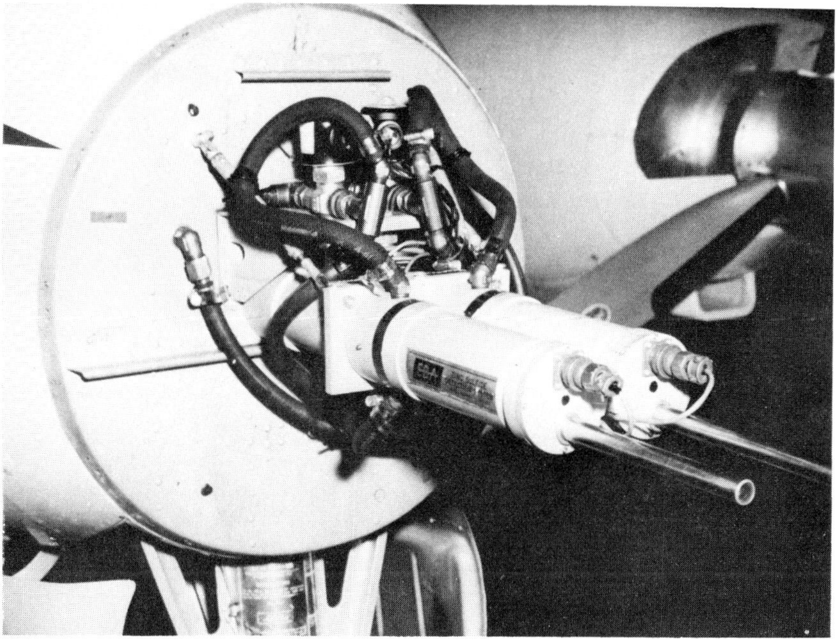

FIG. 2. Closeup of probes shown in Fig. 1. These instruments are used in studying atmospheric diffusion of seeding materials.

FIG. 3. Radar complex in South Dakota. These are surplus military radars converted to weather modification research purposes. They can be used both for recognizing treatable situations and for evaluating the results of treatment.

In many cases, direct-sensing instruments for obtaining microphysical data require a massive platform (such as an aircraft) to get them to the desired location. The presence of the platform often complicates the measurement process. At the present time, however, there is no feasible alternative for the majority of microphysical factors. Fluid and thermodynamical factors are more easily measured by direct-sensing instruments, though not always with the accuracy that is desired. Remote sensing is becoming increasingly important, but the cost of such equipment—for example, a radar—is quite high. The measurement of pertinent air mass factors is the most nearly satisfactory, and these data are routinely available through the U.S. Weather Bureau. Instruments to transmit, process, and display data are rather well developed. Many alternatives are available, each of which is best suited to a particular kind of situation.

C. Treater Function

After recognizing that the situation at hand is treatable, the third activity of an operational precipitation management system is the actual seeding, which first requires selecting the appropriate treatment material and delivery system to achieve the desired result. This involves prior knowledge of the following:

(1) Nucleating characteristics of the various materials available.
(2) Characteristics of the generating or dispersing equipment.

The two most common seeding materials used to induce extra precipitation from cold clouds (clouds that extend vertically to regions where the temperature is lower than 0°C) are, at present, dry ice and silver iodide. Dry ice, because its temperature is $-78°C$, lowers the temperature of the air in its immediate vicinity below $-40°C$. At these very cold temperatures, the water vapor changes phase to become tiny ice embryos, which subsequently act as nuclei for converting supercooled cloud droplets into ice crystals.

Silver iodide particles act as nuclei in the heterogenous nucleation process, but it is not completely clear just how it acts. It was originally selected for experimental work because its crystal structure is very similar to that of ice, the idea being that AgI would "fool" the supercooled droplets into thinking it was the ice crystal needed in the ice crystal precipitation theory (growth from vapor diffusion). There is now evidence [*29*] that an AgI particle is most effective as a nucleating agent when it comes in contact with the surface of a supercooled water droplet (contact nucleation). There is also some evidence [*30*] that collisions between AgI particles and supercooled droplets are rare unless both are electrically charged. In addition, there is some evidence that AgI particles act first as condensation nuclei and then as freezing nuclei.

Some experimentation at the Naval Weapons Center with AgI contaminated with alkali iodides has revealed some interesting results [*31*]. In general, when an alkali iodate (cesium, lithium, potassium, rubidium, and sodium) is included in a pyrotechnic device, the product of the combustion is a silver iodide–alkali iodide complex. These particles are somewhat more effective as ice nuclei than the product of the same pyrotechnic without the alkali.

Regardless of how it works, the threshold of nucleation activity for silver iodide is about $-5°C$. Most other materials that have been tested for their ice-nucleating ability show threshold temperatures lower than $-5°C$. A notable exception is the organic chemical metaldehyde, which becomes effective at about $-1°C$ [*32*]. A number of other organics are effective in the temperature range $0°$ to $-10°C$ (e.g., phloroglucinol and 1,5-dihydroxy-naphthalene), but each has dispensing problems that need to be overcome before it can be considered for operational use.

Warm clouds (clouds whose temperatures are everywhere higher than $0°C$) are generally not affected by materials effective in cold clouds. Common salt (NaCl) is the most frequently used agent to treat warm clouds. Careful sizing of the material is essential to achieve specific modifications of the drop growth spectrum. Shortening the time required to develop precipitation-sized drops is critical in many clouds because of their relatively short life cycle.

The method of producing nucleating material in the form, size, and concentration best suited for precipitation management has received close attention during the last few years. Each seeding material has its unique problems, but this article will consider only a few materials.

Silver iodide particles are most commonly produced by burning, in a propane flame, a solution of silver iodide (usually 2%) dissolved in a solution of sodium or potassium iodide and acetone. Another method burns charcoal or coke impregnated with silver iodide. More recently, pyrotechnic devices producing silver iodide have become readily available. These devices can be made to burn rapidly or slowly, and the amount of AgI can be varied over a wide range with no problems. They can be burned on the ground or in the air; they can be dropped from aircraft flying above the clouds or fired into the clouds from the ground by rockets. An early pyrotechnic device was fired from a Very pistol. They suffer the limitation that once started, they must burn to completion.

The common propane–acetone–silver iodide generator has been found over the years to be a rather dependable piece of equipment. It has been adapted to operate by remote control and is about as reliable as the radio equipment that turns it on and off. In modified form, it can be mounted on an airplane for aerial release of AgI. The greatest handicap of the acetone generator is its inflexibility. Some meteorological situations may require different outputs of AgI than others, but the standard generator is not designed to permit variable output. Recently a variable output generator was designed, and a prototype model constructed [14]. This generator has not yet been used operationally, but it will eventually provide the precipitation manager greater flexibility in treating storms.

FIG. 4. A silver iodide generator located on a mountain ridge in Washington's Cascade Range. The tank on the left contains propane under pressure, and the right tank contains the AgI–acetone solution.

Dry ice is commonly dispersed from the air by placing large chunks of the material into an apparatus, similar in principle to an ice crusher, that breaks the dry ice into more or less uniform pieces from about the size of a pea up to the size of a walnut. These particles are then released to the air, usually at the $-5°$ to $-10°C$ level.

Before salt can be used in cloud seeding it must be ground to a fine powder; the size of the particles is quite important. The salt can be released from the air or, under favorable circumstances, from the ground.

Metaldehyde, which has some excellent nucleation properties, suffers from two deficiencies. Finely ground particles coagulate quickly and also evaporate quickly, making the material unsuitable for release from the ground. Therefore, an airborne metaldehyde generator is required. One model employs a chamber into which metaldehyde powder and the aircraft engine's exhaust gases are fed. The chamber has a butterfly valve on the forward end to admit outside air and thus control the chamber temperature, which must be high enough to vaporize the metaldehyde but not high enough to decompose it. The metaldehyde vapor is immediately released to the free air where it quickly turns to a fine powder and is dispersed by the turbulence [*33*].

Given the information about nuclei and generator characteristics presented above, the precipitation manager must choose a seeding material and a delivery system that will optimize the desired results *as applied to the storm at hand.* To do this, he must have additional information, this time about the transport and diffusion of nuclei under the existing conditions. He must assure himself that (1) the material will arrive at the effective level in the cloud, (2) that the effect will be felt at the ground in the desired area, and (3) that no undesirable effect will occur outside his target area.

Consider for the moment the use of ground generators. Vertical air currents carry the material from the generator to the cloud. The vertical motion can result from the thermal structure of the air or from the mechanical lifting of the air (by a mountain or by a front, for example). If a strong inversion exists above the generator level, the seeding material will not be able to rise above the inversion. If the effective level (the level at which the particular seeding material becomes active as a nucleating agent) is above the inversion, the material will never get the chance to do its job. The seeding will, in effect, be unsuccessful.

If the vertical motion is such that the material can reach the effective level, the precipitation manager is next faced with the questions: (1) Will the concentration of nuclei be proper for optional effects? and (2) will the effects be limited to the desired area on the ground? If the concentration of material reaching the effective level is too great, overseeding can result. In the case of cold clouds, this converts all the water into a large number of very tiny ice particles, too many and too small to stimulate the ice crystal precipitation process. However, some of these ice particles may drift downwind into a region of supercooled droplets where the classical ice crystal theory will operate. Unfortunately, this region may be beyond the target area. If too little material is involved, no significant effect will be observed. As mentioned in an earlier section, the "just right" concentration of artificial nuclei is really the difference between the optimal concentration and the concentration of natural nuclei present.

The horizontal wind speed and direction, plus their variations, determine in large measure where the seeding material will go and where the effects will be felt. A number of diffusion equations exist that can be used to predict the path and the relative concentration of seeding material. The equations usually consider the stability of the air (which can be related to vertical motion) as well as the wind, but none of the equations will yield a perfect solution. Often empirical data, in the form of nomograms, for example, provide better results than formal equations.

Another important consideration in the targeting process is the time required for the artificial nucleus to arrive at the effective level, nucleate a supercooled droplet, grow to precipitation size, and fall to the ground as rain or snow. It is estimated that about 20 to 30 minutes is required for a nucleated droplet to grow to precipitation size. During this time, the particle is moving with the wind at its level. The height at which the droplet is nucleated, the vertical velocity (which may carry the ice crystal up), the horizontal velocity, and the terminal velocity of the falling particle determine where the particle will reach the surface.

Computer models are beginning to appear that can determine the path and concentration of nucleating material from the generator to the clouds, then compute the effect on the cloud of the artificial nucleation and estimate the amount of additional moisture, and finally predict the fallout trajectory of the precipitation. The models are, of course, crude, but they will give the precipitation manager a quantitative estimate of what is likely to happen if he seeds a given storm.

Before leaving the discussion of the ground generator, consider the following complex situation. A silver iodide generator is located on a mountain top far enough from the target area that, under "average" conditions, the results of seeding winter storms will be found in the target area. On a particular day the stability and wind situation are such that part of the AgI goes up into the cloud and part goes down the mountain into the valley below. The AgI going into the cloud diffuses in such a way that some of it reaches the effective level in a matter of a few minutes. The release of latent heat stimulates buoyancy, and the resulting vertical motion carries the tiniest of the newly formed ice crystals higher up in the cloud, while the largest crystals begin to fall out as snow. As more of the AgI particles reach the effective level, somewhat farther downwind than the first, more latent heat is released, and the resultant vertical motion is added to the natural vertical motion. Assuming that the concentration of AgI at the effective level is about optimal, the precipitation process should proceed smoothly and the results should be felt as increased snow in the target area. But what about the AgI that went into the valley? Suppose that during the day thermal convection carries this AgI into the cloud. This could result in overseeding or in expanding the area of effect beyond the target area. In an actual situation, the process is frequently much more complicated than this.

Airborne seeding can eliminate the problem of getting the seeding material to the effective level. However, instead of supplying a continuous supply of nuclei, airborne seeding affects a relatively small volume in any period of time, and that volume is often overseeded. This situation can be overcome by carefully planning the seeding path and need not be considered a serious drawback in most situations. The hazards of flying in mountainous terrain or in heavy icing or turbulent conditions must be considered. More often than not, the best seeding opportunities, especially in winter orographic storms, are the very situations a reasonable pilot would go out of his way to avoid.

The problem of precise targeting is still of major concern. Airborne seeding permits more flexibility than ground-based seeding, as far as the point of release of the seeding material is concerned. The major barrier to effective targeting is the difficulty in predicting the three-dimensional wind field between the generator and the target area.

D. Evaluator Function

The final activity of an operational precipitation management system is evaluation. Evaluation in a broad sense includes (1) monitoring the opportunity recognizer and treater activities to be sure that the original justification for the project (determined by the analyzer function) still exists and to be sure that all conditions (hydrologic, ecological, etc.) remain within predetermined safety limits, and (2) determining the magnitude of the precipitation increase (or whatever the purpose of the project is).

Proper evaluation requires a sound project design. Thus, the first step in the evaluation activity, the project design, must precede both the recognition and treatment activities. The design must include the following:

(1) A clear definition of the project objectives.

(2) Criteria to be used in choosing treatable storms.

(3) The treatment materials and delivery systems that can be used in the project.

(4) The kinds of data that must be collected.

(5) The number and kinds of instruments that must be installed (consideration must also be given to maintenance of instruments and equipment).

(6) The number of people involved in the project.

(7) The kinds of statistical tests that can be applied, including the advantages and disadvantages of each. Nonparametric tests are valuable in weather modification research because many of the data are not exact but are in effect simply ranks, and there is often no reason to assume that the parent population is distributed in a certain way (e.g., normal distribution).

(8) Consideration of such nontechnical factors as public reaction to seeding and effects on local flora and fauna, and legal questions such as licensing requirements, insurance, and access to public and private lands.

The design can be completely objective, that is, a set of instructions that always result in the same actions, or it can be partly subjective, in which the professional skill of the personnel involved can enter in.

Evaluation of the results of cloud seeding will always require comparisons between (1) data from seeded areas and unseeded areas, or (2) data from seeded periods and unseeded periods. Comparisons of data from different areas require an understanding of the degree of similarity of data from the different areas. Generally speaking, the closer two areas are, the greater the degree of similarity. However, the closer the areas the greater the possibility of accidently seeding part of the "unseeded area." On the other hand, comparison of data from a single area for different periods of time requires understanding the time variation of the data. The randomized crossover design is frequently used in weather modification projects because it permits both area and time comparisons.

Precipitation management evaluation is somewhat complicated because the product delivered (precipitation) is often not quite the same as the product desired (streamflow or water in storage, for example). In addition, the point of delivery of the precipitation is often hundreds of miles from the point of consumption. What data should be used to measure the "real" effect of cloud seeding?

Precipitation can conceivably be measured at the base of the cloud, and statistical tests can be applied to determine the percentage attributable to cloud seeding. It is possible, especially in arid and semiarid regions, that much of the precipitation leaving the cloud base will evaporate before reaching the ground. It can be concluded that cloud-base precipitation amount is not an important statistic in an operational program.

Precipitation can be measured at the surface by a wide variety of devices. All but a few give point measurements, and one can always ask how representative of an area is that one point. Studies have been made on the number of rain gages needed to adequately describe the precipitation of an area. The results of one study [*34*] showed that a rain gage density of 1 per 100 square miles gave almost as good results as a density of 14 per 100 square miles.

A portion of all precipitation is rapidly returned to the atmosphere by evaporation from the soil, from leaves, and from ponds, rivers and other bodies of water. Another portion is transpired by vegetation in the area of the precipitation. Still another portion goes into the soil. The remainder goes into streams or lakes by direct runoff or by underground flow of some kind. More often than not, increased streamflow and the subsequent storage of the water are the desired products of cloud seeding. However, streamflow cannot, in general, be used to evaluate the effect of cloud seeding on a storm-by-storm basis. When precipitation is in the form of snow, even monthly runoff values may not be useful. Annual runoff figures give a fairly good indication of the

FIG. 5. A weighing precipitation gage located in Utah's Wasatch Mountains. Immediately below the black gage, is a pressure transducer. The tower on the right holds the antenna for the radio telemetry system that transmits the signal to the terminal readout center.

value of cloud seeding over a particular area, in that they integrate precipitation over the entire area and represent the usable portion of that precipitation. But only one data point per year per stream is available.

Another approach to the problem of evaluating results concerns the amount of seeding material in a sample of precipitation. Recently, one technique for detecting silver in precipitation samples was developed to the point that it can be used operationally [*35*]. In this technique, the silver is concentrated and extracted by ion exchange methods. Neutron activation analysis is used to determine the amount of silver that was present in the precipitation sample. Precipitation from unseeded clouds typically has a silver concentration of about 10^{-11} gm/cc or less, while precipitation from seeded clouds has a concentration an order of magnitude or more higher. The procedure can detect silver from 1-liter samples in concentrations as low as 3×10^{-12} gm/cc.

The presence of silver in a precipitation sample does not necessarily mean that the silver (as silver iodide) took part in the precipitation process; the absence of silver in concentrations greater than background does not mean that seeding with AgI did not affect the precipitation process. Silver can appear in a precipitation sample by (1) having participated in the precipitation process, (2) scavenging by a falling ice crystal or rain drop, or (3) deposition on the surface of a snowpack or in a precipitation gage. On the other hand, silver iodide seeding could be a triggering mechanism, stimulating the natural precipitation process by releasing a large quantity of latent heat in a small volume, which in turn might result in greater vertical development and longer life of the treated cloud.

A new technique, still in the experimental stage, indicates whether the silver in a snowflake is in the center (and presumably is the nucleus) or at the surface of the crystal (scavenged by the falling crystal). The technique is simple and relatively inexpensive. Ice crystal replicas are made on a glass slide coated with Formvar (polyvinyl acetyl resin, usually dissolved in ethylene dichloride). The replica slide is partially immersed in a supersaturated solution of AgI and KI for 2 to 3 hours. The result is a growth of AgI crystals over the ice crystal replicas that are of sufficient size for low-power microscopic examination. Those crystals growing from the center of the replica are quite distinct from those growing from the edge. Knowing what percent of the crystals have silver in the nucleus, what percent of the silver in a representative sample of snowflakes is in the center, and what percent of the silver is scavenged would give added meaning to the total silver content of a precipitation sample.

One of the more significant problems in evaluating precipitation management concerns targeting the effects of seeding. There is evidence that, under some meteorological conditions, seeding affects the dynamics of the atmosphere in such a way as to cause alternate areas of relatively small decreases and substantial increases in precipitation downwind of the primary target area for distances of a hundred miles or more [*14*, *36*]. In at least one instance, the downwind increases exceeded the increase in the target area. There is also some evidence that this dynamic effect extends upwind and laterally as well.

Another targeting problem is nuclei carryover, that is, artificial nuclei carried downwind of the target area. In this situation, low-level nuclei could be carried by the wind near the surface to areas well downwind of the target. Subsequent vertical motion could carry them up into clouds and increase the precipitation where it is not wanted. It is also possible that meteorological conditions over the target could be such that a large fraction of the seeding material would pass unused out of the area.

One of the most persistent criticisms of precipitation management is the idea that increased precipitation in one area necessarily leads to decreased precipitation somewhere else. There are several ways to show that this is a needless fear. Consider the atmospheric water budget described earlier. Only 10% of the water in the atmosphere falls out as precipitation. Precipitation management can probably increase this by 10% to 20% on an annual basis. This would seem to indicate 10% to 20% less moisture in the air. But remember that much of the increased precipitation will be returned to the atmosphere by evaporation and transpiration. Remember also that the atmosphere is three-dimensional and is constantly mixing. Thus, if a given area is seeded and the moisture reduced by, say 50%, much of that moisture deficit would be made up by vertical and lateral mixing in a relatively few miles.

FIG. 6. Aerial view of clouds forming over Colorado's Park Range. Notice the pyrotechnic seeding devices at the lower left.

Next consider precipitation in mountainous terrain. The windward side of the mountain almost always receives more precipitation than the leeward because the air passing over the crest of a mountain tends to flow down the leeward side. This subsidence, and consequent warming, are natural enemies of precipitation. The precipitation gradient on the east side of the Cascades, Sierra, and Rocky Mountains amounts to 20 inches or more in less than 50

FIG. 7. Rain shafts extending below a seeded convective cloud in South Dakota. Observers on the ground measured over 2 inches of rain from this cloud as it passed overhead. The control cloud failed to produce any precipitation.

miles. Cloud seeding cannot worsen this situation, and, in fact, may significantly improve it.

Now consider convective situation, such as thunderstorms. The duration of an average convective cloud is measured in minutes. A major portion of the precipitation that falls came from the area over which the cloud formed. Cloud seeding in this case increases the percentage of moisture that falls, but has little effect on the moisture potential of other areas.

Finally, consider the downwind effects described previously. When the dynamic downwind effect is present, the significant result is precipitation increase, not decrease. Greater understanding of the precise cause of the dynamic effect will almost certainly lead to greater control over it.

V. Outlook for the Future

While a great deal of progress has been made in the last 20 years, precipitation management is not yet the efficient system it needs to be to provide the maximum benefits from atmospheric water resources. Many of the basic techniques used today were developed 15 to 20 years ago, and the major advances have been in learning how to use them more efficiently. Perhaps the most significant progress has been the identification of the critical problem areas that stand in the way of early application of a truly operational precipitation management system [*14*].

The field of weather modification, to date, has been concerned with problems of limited scope and scale. Most of the attention has been devoted to the ice crystal aspects of modification, and it has only recently appeared that dynamic factors are probably of greater importance. Because of the great variety of important problems that have become apparent in weather modification, many of those working in the field have tended to develop their own special fields of interest in increasing depth but with decreasing contact with other workers in allied fields of modification.

The present status in the field can be described as extremely encouraging when attention is given to single cloud experiments or limited area seeding in winter storms. Continued growth in the field will require that the perspective of the program be broadened to include other concepts of modification and to include larger scales of interest. As this perspective is broadened, new problems will arise and solutions to many of the old problems will be found to be less vital than had been originally thought.

Attention should be devoted to concepts of weather modification that serve to redistribute or decrease precipitation by altering microphysical growth rate. Of paramount importance, however, is to learn how to use efficiently the latent heat released by the modification process. Desired products of the latent heat use are increased vertical cloud depth, increased horizontal cloud dimensions, control of the location of the precipitation, and control of the storm movement. There is every reason to believe that the balance of natural forces in cloud systems can be altered much more significantly through the convection process than by simple microphysical changes in hydrometeor size.

Increased progress in weather modification will require increased use of modeling concepts in both computational and laboratory scales. This becomes particularly apparent when larger field programs are contemplated involving large expenditures and large observational capabilities. More attention must be given to quantitative verification of modeling results in terms of actual field data so that sufficient confidence can be gained in the models to extend the concepts to more complex situations. Orderly, efficient progress in the field will come only from a continual interplay between computational, laboratory, and field studies.

The need for larger scale field programs has already become apparent. A desirable concept for the future is an area equipped with a large observational network capable of defining the air motions to the degree required for understanding the environment in which the modification operates. A concentrated effort should be made to define and model these motions through an appreciable range of synoptic storm conditions. Within this observational framework, various workers from different organizations should carry out key experiments on a variety of scales as their interests and needs of the field dictate. This type of program would then evolve toward a prototype operational project as experience in the area and understanding of local effects increase. The key feature required to initiate this type of program on a sound basis is an early and diligent effort aimed at defining the dynamical fields of motions which are present in the area during periods when modification might be contemplated.

There is ample need for certain key experiments on a smaller scale where answers to specific questions are required. Chief among these experiments would be quantitative model verification in the field. Increased complexity of modeling can be accomplished only if various development phases of the model are checked and confidence established in the model assumptions at each stage. Additional small-scale experiments are required to evaluate coalescence models, to relate nuclei counts to the creation of ice crystals in the cloud, and to develop additional understanding of the ice growth mechanism. The increasing use of tracer technology should prove to be a highly valuable tool for these small-scale experiments.

A vital experimental program involves the manipulation of heat to organize convection in the desired manner. The development of this technique would provide an approach toward the creation of precipitation in preferred areas in contrast to the limited capabilities achieved by increasing existing precipitation. Special concepts of modification and special experimental programs will be required for these studies. Emphasis will have to be placed on volume temperature and buoyancy measurements rather than microphysical details. Effects of additives such as carbon black or salt need to be reevaluated as a possible stimulant to the dynamic motions.

With the advent of larger scale interests, there is increased need for modeling prior to the field experiments. Along with the computer modeling, laboratory models are applicable to these scales of interest whereas they are not particularly appropriate to the small-scale cloud motions. Orographic flow and downwind effects should be readily modeled in laboratory tunnels, providing realistic atmospheric parameters are simulated.

It is envisaged that large-scale modification programs of, perhaps, 200 miles downwind extent represent the future goal which weather modification should work toward, at least on an intermediate time scale. Very much greater emphasis on experimental design will be essential for the efficient conduct of these programs. Laboratory and computer modeling can assist in these designs but the controlling factors in the success of the programs will be the clear definition of objectives and goals and techniques for their accomplishment.

Several specific problem areas can be discussed in more detail.

(1) *Microphysics.* There is an urgent need to formulate a microphysics ice phase model in considerably more definitive terms. The model must include a dynamics framework so that microphysics growth rates in various updraft environments can be studied. The model is required to evaluate optimum amounts and locations of seeding material to accomplish various desired objectives. Major factors needed to develop an adequate model are fall velocity and density data for large crystal sizes as a function of temperature and crystal shape, aggregation (ice–ice collection) data, and the influence of electrical effects on various aspects of ice crystal growth. Once the model is developed in more detail than presently available, a series of calculations should be performed covering a range of cloud conditions, seeding amounts, and locations to generate data on optimization techniques for modification.

There are other troublesome areas in ice microphysics that can be considered as extensions of the basic ice phase model. These relate to the nucleation activity in realistic cloud conditions. The importance of drop freezing as a method of generating ice in the cloud should be evaluated and the appropriate conditions for its occurrence defined. The relation of natural nuclei counts or artificial nuclei counts to the subsequent occurrence of ice in the cloud has not been established in any detail. The requirements of extensive natural nuclei background counts will depend largely on establishing a meaningful in-cloud versus outside-cloud relationship. Inherent in this problem is the possibility for ice multiplication in the real cloud system under certain conditions. Recent work has indicated that the dynamics of the cloud system cannot be separated from an evaluation of the ice microphysics growth problem.

In the coalescence area, quantitative evaluations of the existing coalescence models are required. Ideally, these should be carried out under simple conditions where no ice exists so that definitive results can be obtained. These data are required before more complex models and concepts of mixed condensation–ice nuclei modification can be considered.

(2) *Cloud Dynamics.* The one-dimensional cloud models require additional considerations of introducing the ice phase and techniques for parameterizing the mixing effects. These are details that will be solved by additional computer studies and by additional testing in the field. New concepts needed are consideration of the dissipative stage of the cloud and the effects of change in cloud size during its life cycle. Increased effort on quantitative field tests of the models is urgent.

Two-dimensional models offer an opportunity for the testing and evaluation of the importance of numerous secondary cloud parameters. In the near future, the two-dimensional models are not likely to replace the one-dimensional formulations in those cases where computations for a variety of environment and cloud conditions are required. A steady, continuing effort on two-dimensional models is required, however, to look into the importance of environment interactions, wind shear, etc.

The major new dynamics area that is receiving insufficient attention is the modeling of organized cloud systems on a scale larger than individual clouds. Included in the required models are mesoscale and orographic flow, downwind effects together with the generation of populations of clouds. With the apparent trend toward larger scale interests, it is essential that a substantially increased effort in modeling of this scale be undertaken.

(3) *Treatment Potential.* The entire subject of treatment potential has received a disproportionately small emphasis, to date. In part, this is the result of the inability to define potential in meaningful terms. Recent work has begun to make significant inroads into this problem and it is appropriate to increase the effort in this area substantially.

The approach to the problem should be through model computations performed on climatological data but verified by field experiments, as required. Additional observing requirements for the future may very well be a necessity for further delineation of the potential for modification. It is urgent that the framework for the potential evaluation be established as soon as possible so that any additional field experiments or observations can be accumulated at an early date.

The new problems in this area again are associated with the trend toward larger scale programs. The opportunities for mesoscale organization, orographic storm potential, and cyclonic storm modification have not been considered in any detail. New concepts for accomplishing the modification such as continuous movement within the storm over an extended period of time may require new modes of thinking in regard to the modification potential. Regional synoptic studies, including detailed flow characteristics in the area, should be initiated, aimed at development of prototype or operational programs in the probable areas of major interest.

(4) *Instrumentation.* The principal instrumentation problems relate to the acquisition and processing of increasingly larger amounts of data, preferably on a real-time basis. Techniques for accomplishing the desired objectives are, for the most part, available in the modification field already or could be found in the aerospace industry. Reasonably priced, economical systems for accomplishing the desired ends are much more difficult to obtain. A considerable effort may be required to define the needs of the weather modification field as clearly as possible so that reasonable simplifications in the system can be carried out. The essential features required of the system are the real-time display of dynamic and microphysics information under storm conditions.

Specific instrumentation problems of a new or improved nature include snow measurement, the use of Doppler radar (vertical or multiple systems), remote probing, and measurement of mesoscale wind fields. Increased utilization of high quality airborne instrument platforms is a necessity for further research studies. The capabilities for utilizing satellite data in modification programs, particularly synchronous satellites, have been little explored.

(5) *Warm Cloud Modification.* A detailed field test of the warm cloud coalescence model is urgently needed. This can be done relatively simply with condensation nuclei, drop size measurements and calibrated radar. Effects of nuclei seeding should also be investigated in the tests in order to see if the model adequately predicts the treatment of effects.

There is a dearth of information on the distribution of naturally occurring condensation nuclei. This is true both for the small nuclei which determine cloud droplet concentrations and the large ones which directly serve as rain drop embryos. Since the rainfall modification techniques involve adding to these nuclei, information on the background concentration is vital for planning modification techniques and conducting experiments. Knowledge of the chemical composition and sources of these nuclei should be improved, and there needs to be much more data on the background levels versus location and season. Of particular practical importance is the alteration of the nuclei situation with distance inland. Present evidence suggests that, in the absence of pollutant sources, typical marine values may prevail for some distance inland. This possibility needs to be clarified by an observational program, and the downwind effects of various pollutant sources also should be investigated.

There is an enormous lack of information on cloud populations, i.e., number and size distributions of clouds as a function of location and synoptic conditions. Many more data on these cloud populations are needed before a realistic appraisal of the potential for warm cloud modification can be made.

Further consideration should be given to the possibilities of creating dynamic effects in clouds in addition to the microphysical aspects which have been emphasized in this report. The direct dynamic effects from the release of heat or materials appear to be small. Only the use of carbon black appears potentially capable of adding enough buoyancy to alter the cloud situation in any significant manner.

(6) *Legal, Economic, Ecological, and Sociological Considerations.* The atmosphere and its moisture cannot be considered to be independent natural resources. The interactions and interdependencies among the atmosphere, the soil, the oceans, plants, animals, and man clearly point out the necessity to consider atmospheric water as an integral part of our total natural resource.

A special study of the legal implications of precipitation management [*37*] examined current laws and legal theories concerning water rights, liability for damages, and regulations concerning weather modification operations. There is growing awareness that present State and Federal laws are generally inadequate for the optimal development and utilization of precipitation management technology.

The economic consequences of the application of precipitation management technology are difficult to assess, and little more than cursory analyses have been made thus far. One of the most difficult problems lies in determining the value of each additional increment of water. Historically, water has been considered a free economic good or an integral part of a piece of land. More research is needed to define the economic implications of weather modification.

A change in the precipitation regime of an area is bound to have an effect on the plant and animal communities of the area. In some situations, the change will be for the better, but in other situations the balance of nature could be upset and an undesirable condition could result. Research into the ecological consequences of precipitation management has not received the attention it deserves. While the development of additional water is necessary, almost certainly overshadowing in importance the detrimental effects on a

FIG. 8. Precipitation falling from that portion of a convective cloud seeded with a pyrotechnic flare. No precipitation was observed from the unseeded portion.

few plants or animals, the full extent of those effects should be understood. The establishment of ecological bench mark stations has been suggested as one way to provide the necessary information.

The sociological implications of precipitation management are most complex. Each subgroup of the population has its special reasons to support or oppose weather modification. Often resistance to a precipitation management project disappears, or at least decreases, when the facts are presented in an unemotional atmosphere. Research is needed to understand the reaction of social groups to the advent of precipitation management.

One of the most noticeable changes in the immediate future will be the movement toward large-scale field projects designed to test precipitation management systems in an operational mode. The purpose of these projects is not to develop new techniques but to evaluate the effectiveness and efficiency of a given system in a particular location. The desirability of early implementation of precipitation modification technology is obvious; so also is the desirability of "calibrating" that technology. The transition from purely experimental projects to purely operational projects should not be expected to proceed without problems. The pilot-plant concept can be applied effectively to precipitation management. It must be demonstrated that an operational precipitation management system will operate properly, both scientifically and economically. Perhaps most important, it must be demonstrated that the system can be integrated into the overall natural resource development of the region.

References

1. The Nation's water resources, summary rept. Water Resources Council, Wash., D.C., (1968).
2. Final rept. of the Advisory Committee on Weather Control, Vol. I: Summary of findings and recommendations, and Vol. II: Tech. repts. U.S. Congress, Wash., D.C. (1957).
3. Weather and climate modification problems and prospects, Vol. I: Summary and recommendations, NAS-NRC Publ. No. 1350, Final rept. of the Panel on Weather and Climate Modification to the Committee on Atmospheric Sciences. Nat. Acad. of Sci—Natl. Res. Council, Wash., D.C. (1966).
4. Ninth ann. rept. on weather modification (Fiscal year 1967), NSF 68-21. Natl. Sci. Foundation, 78–87 (1968).
5. Battan, L. J., Silver-iodide seeding and rainfall from convective clouds. *J. Appl. Meteor.* **5**, 669–683 (1966).
6. Battan, L. J., Silver-iodide seeding and precipitation initiation in convective clouds. *J. Appl. Meteor.* **6**, 317–322 (1967).
7. Braham, R. R., Jr., Project Whitetop—A five-year randomized cloud seeding study. Paper presented at the 24th Natl. Meeting of the *Am. Meteor. Soc.* on Cloud Physics and Severe Local Storms, Reno, Nevada, Oct. 18–22, 1965.

8. DECKER, W. L., and SCHICKEDANZ, P. T., Final report on Project Whitetop; Part IV—A summary of the rainfall analyses, Report to Natl. Sci. Foundation, Atmospheric Sci. College of Agriculture, Univ. of Missouri, 1966.
9. State of California, Santa Barbara Weather Modification Project—Interim Rept. of the Board of Directors. Dept. of Water Resources, State of California, 1960.
10. GRANT, L. O., and MIELKE, P. W., Jr., A randomized cloud seeding experiment at Climax, Colorado, 1960–65, *Proc. 5th Berkeley Symp. on Math. Statist. and Probability*, Vol. V: Weather modification experiments. Univ. of Calif. Press, 115–131, 1967.
11. GRANT, L. O., and CHAPPELL, C., unpublished manuscript, 1969.
12. EBERLY, D. L., and ROBINSON, L. H., Design and evaluation of randomized wintertime cloud seeding at high elevation. *Proc. Berkeley Symp. on Math. Statist. and Probability* 5th, Weather Modification experiments. Univ. of Calif. Press, **5**, 65–90, (1967).
13. DENNIS, A. S., and KOSCIELSKI, A., Results of a randomized cloud seeding experiment in South Dakota, *J. Appl. Meteor.* **8**, 556–565 (1969).
14. U.S. Dept. of the Interior, Project Skywater 1968 Ann. Rept. Vol. I, Summary; Vol. II, Contractor rept. Bur. of Reclamation, Office of Atmospheric Water Resources, (1969).
15. CALVIN, L. D., Statistical analysis of the Elko weather modification experiment, 1963–1968. Tech. Rept. No. 9, Interim Rept. for Project Skywater, Laboratory of Atmospheric Phys., Desert Res. Inst., Reno, Nevada, (1969).
16. SMITH, E. J., ADDERLEY, E. E., and WALSH, D. T., A cloud-seeding experiment in the Snowy Mountains, Australia, *J. Appl. Meteor.*, **2**, 324–332 (1963).
17. GABRIEL, K. R., The Israeli artificial rainfall stimulation experiment; statistical evaluation for the period 1961–65. *Proc. Berkeley Symp. on Mathe, Statist. and Probability 5th* Weather modification experiments. Univ. of Calif. Press, **5**, 91–113 (1967).
18. GABRIEL, K. R., Recent results of the Israeli artificial rainfall stimulation experiment. *J. Appl. Meteor.* **6**, 437–438 (1967).
19. BISWAS, K. R., KAPOOR, R. K., KANUGA, K. K., and RAMANA, MURTY, Bh. V., Cloud seeding experiment using common salt. *J. Appl. Meteor.*, **6**, (5) 914–923 (1967).
20. HENDERSON, T. J., Cloud seeding on the Kings River Watershed, Final rept. on the 1966–67 operations. Atmospherics, Inc., Fresno, Calif. (1968).
21. ELLIOTT, R. D. and LANG, W. A., Weather modification in the southern Sierras, *J. Irrigation and Drainage Div.*, ASCE, **93**, No. 1R4, 45–59 (1967).
22. BYERS, H. R., " Elements of Cloud Physics." Univ. of Chicago Press, Chicago, Illinois. 1965.
23. FLETCHER, N. H., " The Physics of Rainclouds." Cambridge Univ. Press, London and New York, 1962.
24. MASON, B. J., " The Physics of Clouds." Oxford Univ. Press, London and New York 1957.
25. FUKUTA, N., Unpublished rept. to the Office of Atmospheric Water Resources, U.S. Bur. of Reclamation, Denver, Colorado, (1968).
26. MORGAN, G. M., and ALLEE, P. A., The production of potential ice nuclei by gasoline engines. *J. Appl. Meteor.* **7**, 241–246 (1968).
27. SCHAEFER, V. J., Ice nuclei from automobile exhaust and iodine vapor. *Science* **154**, 1555–1557 (1966).
28. SCHAEFER, V. J., New field evidence of inadvertent modification of the atmosphere. *Proc. Natl. Conf. Weather Modification 1st Am. Meteor. Soc.* 163–172 (1968).
29. GOKHALE, N. R., and GOOLD, J., Jr., Droplet freezing by surface nucleation, *J. Appl. Meteor.* **7**, 870–874 (1968).

30. Berg, T. G. O., Nucleation and growth in cloudseeding. *Proc. Skywater Conf. I, Phys. Chem. Nucl.* U.S. Dept. of the Interior, Bur. of Reclamation, Denver, Colorado, 127–146 (1967).
31. Finnegan, W. G., Chemical and physical properties of freezing nuclei from pyrotechnic combustion. *Proc. Skywater Conf. I, Phys. Chem. Nucl.* U.S. Dept. of the Interior, Bur. of Reclamation, Denver, Colorado, 246–284 (1967).
32. Fukuta, N., Ice nucleation by metaldehyde, *Nature*, **199**, 475–476 (1963).
33. Fukuta, N., An airborne generator of metaldehyde smoke. *J. Appl. Meteor.* **6**, 948–951 (1967).
34. Nicks, A. D., Field evaluation of rain gage network design principles. WMO-IASH Symp. on Design of Hydrometeorlogical Network, Quebec City, June 15–22 (1965).
35. Warburton, J. A., and Young, L. G., Nuetron activation procedures for silver analysis in precipitation. *J. Appl. Meteor.* 7, 433–443 (1968).
36. Adderley, E. E., Rainfall increases downwind from cloud seeding projects in Australia. *Proc. Natl. Conf. Weather Modification, 1st Am. Meteor Soc.* 42–46 (1968).
37. The legal implications of atmospheric water resources development and management, Rept. to U.S. Bur. of Reclamation, Contract No. 14-06-D-6224. College of Law, Univ. of Arizona, Tucson, Arizona (1968).

THEORY OF FLOW TO DRAINS

JAN VAN SCHILFGAARDE

Soil and Water Conservation Research Division
Agricultural Research Service
U.S. Department of Agriculture
Beltsville, Maryland.

I. Introduction

In his 1967 State of the Union Message, President Johnson said:

> Next to the pursuit of peace, the really greatest challenge to the human family is the race between food supply and population increase. That race tonight is being lost.

The report of the Panel on World Food Supply of the President's Science Advisory Committee [*1*] pointed out that, in Asia, the need for increased food production can be met only by increasing the production per acre through the development of the substantial water resources available. The total cost of water resource development for agriculture in the Indian subcontinent and in Southeast and Southwest Asia was estimated at $80 billion. In other areas, the report also stressed the need for better water management.

By itself, improved water management cannot meet the need for increased production, or production efficiency, in agriculture. In concert with other biological, engineering, and economic technologies, however, water management takes a place of prominence among the inputs that promise high return.

The term "water management" implies partial or total control of the water on the land and in the soil to enhance a high level of production and reduce damages to a minimum. One phase of this control involves land drainage. The economic importance of drainage may be illustrated by the following statistic. In 1965, the Soil Conservation Service of the U.S. Department of Agriculture assisted in the design and construction of 37,071 miles of open and closed drains [*2*].

The objective of agricultural drainage is the orderly removal of water that is excess to economic crop production or, in the case of irrigated land in arid climates, the establishment and maintenance of a proper salt balance in the soil. The mechanism whereby this objective is obtained involves the physics of flow of fluids through the soil, and consequently depends heavily on the properties of the soil. Installations that will attain the objective require engineering design and construction. Thus we see that drainage implies the attainment of essentially agronomic objectives through the engineering application of physical principles.

This article restates the basic principles on which the theory of fluid flow is based, reviews the developments from these principles as they apply to drainage, and illustrates their application to practical design problems.

At first glance, the problem facing the practicing drainage engineer is so complex that it appears to defy subjugation to idealized theoretical analysis. Surface topography more often is irregular than level or plane. Measurements of hydraulic conductivity on soil samples will show wide variations: hundredfold differences are common even within a small area. Aside from spatial variations in soil properties, any property may vary with time as the result of physical, chemical and biological forces. Impervious boundaries are hard to identify in the field. The installation of subsurface drains is likely to result in inhomogeneity of the soil in the neighborhood of the drain with possible drastic effects on system performance. Couple these uncertainties with the irregular nature of precipitation patterns and the difficulty of prescribing drainage requirements quantitatively, and it becomes understandable that drainage design is still treated as an art more often than as a science.

Yet a strong case can be made for practical application of drainage theory. Aside from the obvious advantage of enhancing better qualitative judgment through understanding of principles, the use of rational design methods leads to identification of specific sources of uncertainty. As technology improves in any one phase of the complex agro-physico-engineering system called drainage, this new information can be readily applied in better design. A greatly improved method for measuring effective hydraulic conductivity of a field is of no use if drain spacing is based on a rule-of-thumb. A better evaluation of the effect on crop yield of water table fluctuations at specified stages of growth can only be put to practical use if the design equations permit prediction of the frequency distribution of water table heights.

No attempt is made at completeness in this review. Certain topics, such as anistropy and nonhomogeneity, are omitted entirely. Others, such as unsaturated flow, are treated lightly. The objective is to present sufficient continuity in reasoning and diversity in approach to enhance an appreciation for the usefulness as well as the limitations of drainage theory in application. The concluding section attempts to integrate the theory into the broader problem of engineering design. Unfortunately, present engineering practice frequently ignores the advances in theory of the last twenty-odd years and it is hoped that this review may stimulate some imaginative schemes for ameliorating this situation. It also may be helpful in the always difficult task of assigning priorities for further drainage research.

II. Equations of Motion

The flow of fluids through porous media is a specialized case in hydrodynamics which, in turn, may be considered as a special branch of continuum mechanics. From mechanics, we learn that to describe the motion of a fluid we require a kinematic, a dynamic and a thermodynamic relation. The first of these leads to a statement of conservation of mass; the second may be derived from a momentum balance; the third takes the form of an equation of state relating the fluid's pressure, temperature and density. The classical equations of motion for viscous fluids, derived in texts on hydrodynamics [*3*, *4*], are the Navier–Stokes equations. Since the viscous drag cannot be neglected in the flow through porous media, these are the equations applicable to ground-water flow. As will be shown, however, the Navier–Stokes equations do not lend themselves to solution in very many situations of practical interest. Hence we are forced to search for a different, simpler framework. To this end, we shall replace the momentum balance with another dynamic relation, derived from experimental observations—Darcy's law.

A. Navier-Stokes Equations

Notwithstanding the practical limitations of the Navier–Stokes equations, their derivation is presented here to illustrate clearly the relation between classical hydrodynamics and ground-water theory. Also, the derivations given in most texts are difficult to follow for practicing engineers; it is hoped that the treatment presented here may clarify the physical significance of the Navier–Stokes equations.

Traditionally, we consider a small element of fluid. This element is chosen to be large enough so that molecular motions may be ignored, but small enough so that it may be considered as a homogeneous "point." There are two parallel ways open to us. We may consider a moving element of fluid, or we may focus attention on an element of space through which fluid moves. The first is called the *Lagrangian method*, which uses material (or moving) coordinates; the second, the *Eulerian method*, uses spatial (or fixed) coordinates. In this article, spatial coordinates are used.

Before deriving a dynamic equation from a momentum balance, it is convenient to express the components of stress on an arbitrarily oriented plane in terms of the stresses in the principal planes. Figure 1 represents a

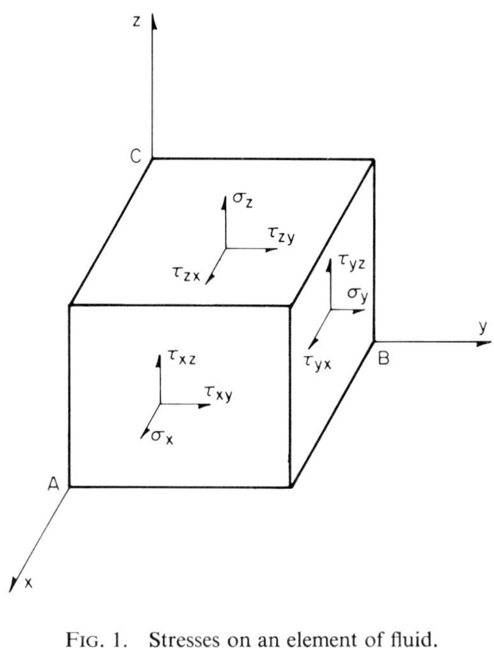

FIG. 1. Stresses on an element of fluid.

small parallelepiped of a medium and shows the three components of stress on each of three faces. The stress on the opposite faces is equal in magnitude but oppositely directed. The stresses on the plane $\overline{ABC}$ in the direction of the coordinate axes, designated by s_x, s_y, and s_z, can readily be obtained from a balance of forces on the prism ABC in terms of the direction cosines[1] l, m, and n:

$$\begin{aligned} s_x &= l\sigma_x + m\tau_{yx} + n\tau_{zx} \\ s_y &= l\tau_{xy} + m\sigma_y + n\tau_{zy} \\ s_z &= l\tau_{xz} + m\tau_{yz} + n\sigma_z \end{aligned} \tag{1}$$

We now consider the rate of change of momentum of a mass of fluid of volume V enclosed by a surface S. Restricting ourselves first to the x direction, the rate of change of the component of momentum parallel to x of the fluid contained in S must equal the sum of (1) the rate at which momentum parallel to x flows into S across the boundary, (2) the rate at which momentum is generated by the impressed body force parallel to x and (3) the component parallel to x of the stresses exerted by the surrounding fluid upon the boundary of S. Designating the fluid density by ρ, its velocity by $\mathbf{v} = \mathbf{i}u + \mathbf{j}v + \mathbf{k}w$ and the body force per unit mass by $\mathbf{i}X + \mathbf{j}Y + \mathbf{k}Z$, the above statement can be written as

$$\int_V (\partial/\partial t)(\rho u)\, dx\, dy\, dz = -\int_S \rho u \mathbf{v} \cdot \mathbf{n}\, da + \int_V \rho X\, dx\, dy\, dz + \int_S s_x\, da$$

Applying Green's theorem[2] and the first of the equations in (1), the surface integrals can be changed to volume integrals. Then, since the above equation must hold for any V, the integrands must meet the same identity as the integrals, so that

$$\frac{\partial(\rho u)}{\partial t} = -\frac{\partial(\rho u^2)}{\partial x} - \frac{\partial(\rho uv)}{\partial y} - \frac{\partial(\rho uw)}{\partial z} + \rho X + \frac{\partial \sigma_x}{\partial x} + \frac{\partial \tau_{yx}}{\partial y} + \frac{\partial \tau_{zx}}{\partial z} \tag{2}$$

Clearly, two parallel equations can be written for the y and z directions. Equation (2) represents the dynamic relation sought.

A mass balance is obtained by considering the rates of mass flow into each of the six faces of the element in Fig. 1, and equating their sum to the rate of change of mass in the element. The result is

$$\frac{\partial \rho}{\partial t} + \frac{\partial(\rho u)}{\partial x} + \frac{\partial(\rho v)}{\partial y} + \frac{\partial(\rho w)}{\partial z} = 0 \tag{3}$$

---

[1] The direction cosines are the cosines of the angles between the perpendicular to the plane $\overline{ABC}$ and the coordinate axes.

[2] $\int_V \left(\frac{\partial f_1}{\partial x} + \frac{\partial f_2}{\partial y} + \frac{\partial f_3}{\partial z}\right) dV = \int_S (lf_1 + mf_2 + nf_3)\, da.$

Expanding the left-hand term and the first three terms on the right of Eq. (2) and utilizing Eq. (3), there results

$$\frac{\partial u}{\partial t} + u\frac{\partial u}{\partial x} + v\frac{\partial u}{\partial y} + w\frac{\partial u}{\partial z} = X + \frac{1}{\rho}\frac{\partial \sigma_x}{\partial x} + \frac{1}{\rho}\frac{\partial \tau_{yx}}{\partial y} + \frac{1}{\rho}\frac{\partial \tau_{zx}}{\partial z}$$

In the left-hand side, we recognize the total or material time derivative of ***u***, Du/Dt. Thus, the result can be written

$$\begin{aligned} \rho\frac{Du}{Dt} &= \rho X + \frac{\partial \sigma_x}{\partial x} + \frac{\partial \tau_{yx}}{\partial y} + \frac{\partial \tau_{zx}}{\partial z} \\ \rho\frac{Dv}{Dt} &= \rho Y + \frac{\partial \tau_{xy}}{\partial x} + \frac{\partial \sigma_y}{\partial y} + \frac{\partial \tau_{zy}}{\partial z} \\ \rho\frac{Dw}{Dt} &= \rho Z + \frac{\partial \tau_{xz}}{\partial x} + \frac{\partial \tau_{yz}}{\partial y} + \frac{\partial \sigma_z}{\partial z} \end{aligned} \tag{4}$$

Equations (4) contain nine different stresses. By considering the rate of increase of the moment of momentum about the axes in a manner analogous to the foregoing momentum balance, it can be shown, by a straightforward, if lengthy, manipulation, that $\tau_{xy} = \tau_{yx}$, $\tau_{xz} = \tau_{zx}$, and $\tau_{yz} = \tau_{zy}$. Thus the number of stress components is reduced to six.

The next step is to relate the stresses to the velocity, or the rate of deformation of the fluid element. To this end, we introduce Newton's law of viscosity, which states that the shear stress is proportional to the rate of change of deformation. Consider (Fig. 2), following [5], a plate *B* that moves

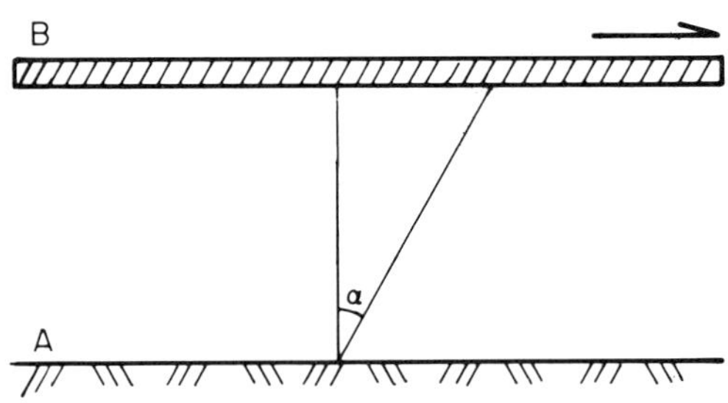

FIG. 2. The concept of viscosity.

parallel to a fixed plate *A*, and let α be the amount by which an originally right angle is diminished as the fluid moves. Then, according to Newton's law,

$$\tau = \mu\, \partial u/\partial y$$

Since $u = \partial x/\partial t$,

$$\tau = \mu\frac{\partial u}{\partial y} = \mu\frac{\partial^2 x}{\partial y\, \partial t} = \mu\frac{\partial}{\partial t}\left(\frac{\partial x}{\partial y}\right) = \mu\frac{\partial}{\partial t}(\tan\alpha) \tag{5}$$

Returning to the element of fluid of Fig. 1, we consider its deformation in the z plane resulting from the tangential stresses τ_{xy} and τ_{yx}, as illustrated in Fig. 3. Making use of Eq. (5),

$$\tau = \tau_{xy} = \tau_{yx} = \mu \frac{\partial}{\partial t} (\tan \alpha_1 + \tan \alpha_2) = \mu \left(\frac{\partial u}{\partial y} + \frac{\partial v}{\partial x} \right) \tag{6}$$

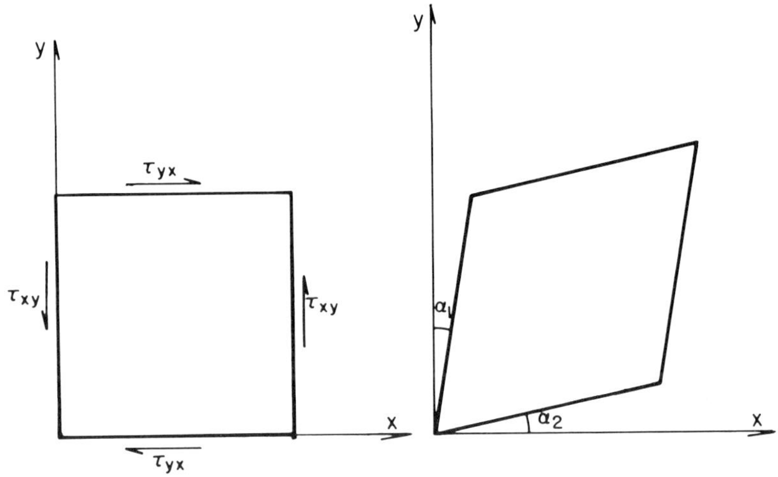

FIG. 3. Deformation of an element of fluid.

Similarly,

$$\tau_{xz} = \tau_{zx} = \mu \left(\frac{\partial u}{\partial z} + \frac{\partial w}{\partial x} \right) \quad \text{and} \quad \tau_{yz} = \tau_{zy} = \mu \left(\frac{\partial v}{\partial z} + \frac{\partial w}{\partial y} \right) \tag{6a}$$

Having found the relationship between the τ's and the velocity components, we next look for a similar relationship for the σ's. As illustrated in Fig. 4, designate the normal stresses on planes rotated by $\pi/4$ from the x and y planes by σ_1 and σ_2. Then, from a force balance on the triangles,

$$\tau_{xy} \sin(\pi/4) + \tau_{yx} \sin(\pi/4) = \sigma_1 \sqrt{2}$$

$$\tau_{xy} = \tau_{yx} = \sigma_1 \qquad \tau_{xy} = \tau_{yx} = -\sigma_2$$

The sign results from the convention of calling compressive stress negative and tensile stress positive. Using Eq. (6), there results

$$\sigma_1 - \sigma_2 = 2\mu(\partial u/\partial y + \partial v/\partial x)$$

The objective, however, is to relate the normal stresses σ_x, σ_y, σ_z, in the same planes as the coordinate axes x, y, z, to the velocity components u, v, w.

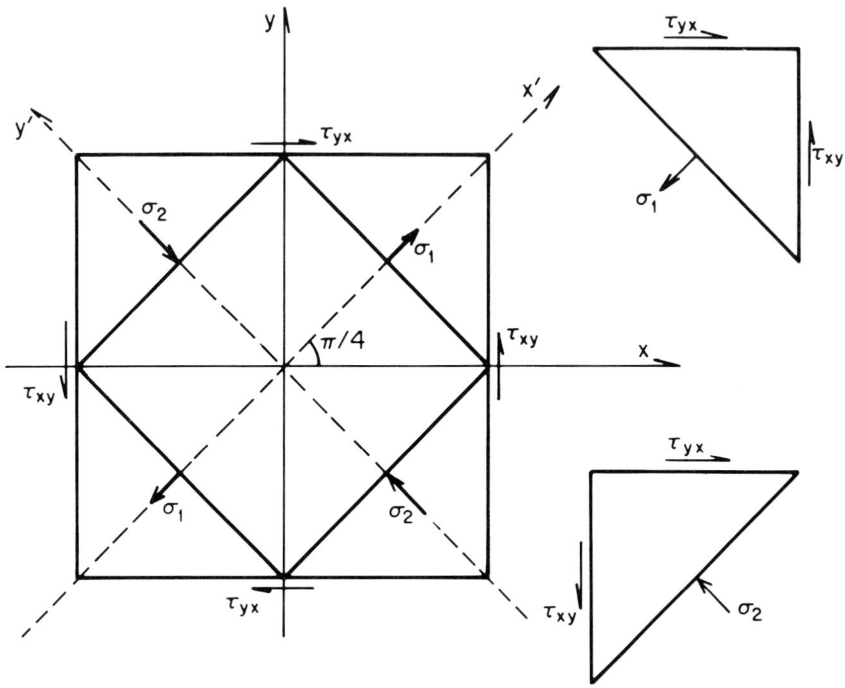

FIG. 4. Relation between tangential and normal stresses.

This is accomplished by a rotation of the coordinate axes through $\pi/4$, yielding

$$x' = (x + y)/\sqrt{2}, \qquad y' = (y - x)/\sqrt{2}$$

In the transformed system, $\sigma_1 = \sigma_x'$ and $\sigma_2 = \sigma_y'$, so that

$$\sigma_1 - \sigma_2 = \sigma_x' - \sigma_y' = 2\tau_{xy} = 2\mu\left(\frac{\partial u}{\partial y} + \frac{\partial v}{\partial x}\right) \tag{7}$$

Since apparently

$$u = (u' - v')/\sqrt{2}$$
$$v = (u' + v')/\sqrt{2}$$

we find

$$\frac{\partial u}{\partial y} = \frac{1}{2}\left(\frac{\partial u'}{\partial x'} + \frac{\partial u'}{\partial y'} - \frac{\partial v'}{\partial x'} - \frac{\partial v'}{\partial y'}\right)$$

$$\frac{\partial v}{\partial x} = \frac{1}{2}\left(\frac{\partial u'}{\partial x'} - \frac{\partial u'}{\partial y'} + \frac{\partial v'}{\partial x'} - \frac{\partial v'}{\partial y'}\right)$$

By substitution into Eq. (7),

$$\sigma_x{}' - \sigma_y{}' = 2\mu\left(\frac{\partial u'}{\partial x'} - \frac{\partial v'}{\partial y'}\right)$$

Since the orientation of the axes is arbitrary, it follows that

$$\sigma_x - \sigma_y = 2\mu\left(\frac{\partial u}{\partial x} - \frac{\partial v}{\partial y}\right)$$

$$\sigma_x - \sigma_z - 2\mu\left(\frac{\partial u}{\partial x} - \frac{\partial w}{\partial z}\right)$$

We now define the fluid pressure p as the mean value of the normal stress on a sphere of unit radius, which yields

$$p = -\tfrac{1}{3}(\sigma_x + \sigma_y + \sigma_z)$$

From the last three equations, we obtain by addition and subtraction

$$\begin{aligned}
\sigma_x &= -p - \tfrac{2}{3}\mu\left(\frac{\partial u}{\partial x} + \frac{\partial v}{\partial y} + \frac{\partial w}{\partial z}\right) + 2\mu\frac{\partial u}{\partial x}\\
\sigma_y &= -p - \tfrac{2}{3}\mu\left(\frac{\partial u}{\partial x} + \frac{\partial v}{\partial y} + \frac{\partial w}{\partial z}\right) + 2\mu\frac{\partial v}{\partial y}\\
\sigma_z &= -p - \tfrac{2}{3}\mu\left(\frac{\partial u}{\partial x} + \frac{\partial v}{\partial y} + \frac{\partial w}{\partial z}\right) + 2\mu\frac{\partial w}{\partial z}
\end{aligned} \tag{8}$$

Substitution of Eqs. (6) and (8) into Eqs. (4) gives the desired result:

$$\begin{aligned}
\rho\frac{Du}{Dt} &= \rho X - \frac{\partial p}{\partial x} + \tfrac{1}{3}\mu\frac{\partial}{\partial x}\left(\frac{\partial u}{\partial x} + \frac{\partial v}{\partial y} + \frac{\partial w}{\partial z}\right) + \mu\left(\frac{\partial^2 u}{\partial x^2} + \frac{\partial^2 u}{\partial y^2} + \frac{\partial^2 u}{\partial z^2}\right)\\
\rho\frac{Dv}{Dt} &= \rho Y - \frac{\partial p}{\partial y} + \tfrac{1}{3}\mu\frac{\partial}{\partial y}\left(\frac{\partial u}{\partial x} + \frac{\partial v}{\partial y} + \frac{\partial w}{\partial z}\right) + \mu\left(\frac{\partial^2 v}{\partial x^2} + \frac{\partial^2 v}{\partial y^2} + \frac{\partial^2 v}{\partial z^2}\right)\\
\rho\frac{Dw}{Dt} &= \rho Z - \frac{\partial p}{\partial z} + \tfrac{1}{3}\mu\frac{\partial}{\partial z}\left(\frac{\partial u}{\partial x} + \frac{\partial v}{\partial y} + \frac{\partial w}{\partial z}\right) + \mu\left(\frac{\partial^2 w}{\partial x^2} + \frac{\partial^2 w}{\partial y^2} + \frac{\partial^2 w}{\partial z^2}\right)
\end{aligned} \tag{9}$$

In vector notation, these equations reduce to

$$D\mathbf{v}/Dt = \mathbf{g} - (1/\rho)\,\nabla p + \tfrac{1}{3}(\mu/\rho)\,\nabla\,\nabla\cdot\mathbf{v} + (\mu/\rho)\,\nabla\cdot\nabla\mathbf{v} \tag{9a}$$

where the body force has been taken as gravity.

Equations (9) or (9a) are known as the *Navier–Stokes equations*. They describe the flow of a compressible viscous fluid. They are exact in the sense that they were derived from the momentum balance, with the assumption that Newton's law of viscosity is valid. They constitute, however, a complicated set of equations that are difficult to solve even for the simplest boundary conditions.

A complete description still requires an equation of state. For incompressible flow, this takes the form $\rho =$ constant, so that, from Eq. (3), the divergence of the velocity vanishes. This causes the third term on the right of Eq. (9*a*), which represents the force associated with fluid expansion, to go to zero. For ideal fluids, where $\mu = 0$, both of the friction terms vanish. The Navier-Stokes equations then reduce to the Euler equations.

The solution of the Navier–Stokes equations requires the identification of appropriate initial and boundary conditions. For flow through porous media, this involves the specification of the exact geometry of the pore space. When the porous medium consists of natural soil or rock, this can never be accomplished except possibly in a statistical sense. Hence practical considerations force us to search for a different, less complicated relation. To this end, we substitute for the momentum balance Darcy's law which is based on a reasonable generalization of an experimentally derived relation.

B. Darcy's Law

In 1856, Darcy [*6*] observed that the discharge rate through a water-saturated vertical column of sand was proportional to the cross-sectional area and the difference in hydraulic head between the top and bottom of the column, and inversely proportional to the column length. The proportionality constant, he wrote, "depends on the permeability of the sand." Thus, in its original form, Darcy's law refers to the flow of water through a vertical sand column and states

$$Q = KA(h_2 - h_1)/L \tag{10}$$

where Q represents the discharge rate, K a coefficient, A the column cross-sectional area, L the column length, and h_2 and h_1 the hydraulic heads at the two termini.

Subsequent experimentation has verified that the equation in the form of Eq. (10) does not depend on orientation, but is valid for horizontal or inclined columns as well. The generalization of Darcy's law to its differential (and three-dimensional) form,

$$\mathbf{q} = -K\,\nabla h \tag{11}$$

has been questioned by some [*7*], but will here be accepted as having proved useful and not at variance with experimental observations.

Darcy's law leads to the consideration of a fictitious velocity, often called the *discharge velocity*, defined as $\mathbf{q} = Q/A$ in the direction of the hydraulic gradient, in contrast to the actual fluid velocity $\mathbf{v}$ of the Navier–Stokes equation. It is this shift from the actual velocity in each pore to the flux across a unit area of the medium that accounts for the greater practical usefulness of Darcy's law.

Dimensional considerations together with experimental observations lead to the relation

$$K = c\rho g d^2/\mu$$

where c is a dimensionless constant related to the geometric pore space arrangement and d some equivalent grain diameter. It has become customary [6] to combine the factors c and d^2 into one and to designate $k = cd^2$ as the *permeability* and K as the *hydraulic conductivity*. The permeability is intended to represent a property of the solid medium, while the hydraulic conductivity is a property of the solid-fluid system. The dimensions of k are apparently $[L^2]$ and those of K are $[L/T]$. Units for k that are frequently used include square centimeters and square microns; for K, one often encounters the units centimeters per second, inches per hour, and feet per day.

Whereas the above distinction between intrinsic permeability of the solid medium and hydraulic conductivity of the solid-fluid system is valid for inert systems, it has been pointed out [8] that in soils containing clay the nature of the fluid may change the geometry of the pore space, so that the concept of a permeability independent of the fluid no longer can be maintained.

Darcy's law has the form of a flux proportional to a driving force, analogous to Ohm's law in electricity and Fourier's law of heat conduction. If we express this driving force in terms of a potential gradient and define the potential of the fluid as the work required to move a unit mass of fluid from a reference state to the point in question, then we note that the potential is given by gh and the driving force by grad gh. This leads to Darcy's law in the familiar form

$$\mathbf{q} = (k\rho/\mu)\text{grad } gh \tag{12a}$$

It is sometimes convenient to use the alternate definitions of the potential in terms of work per unit volume or work per unit weight. The corresponding forms of Darcy's law are, respectively,

$$\mathbf{q} = (k/\mu)\text{grad } \rho g h \tag{12b}$$

and

$$\mathbf{q} = (k\rho g/\mu)\text{grad } h \tag{12c}$$

It should be pointed out that (12b) is not equivalent to (12a) and (12c) unless ρ is constant.

This discrepancy between Eqs. (12a) and (12b) calls attention to an important limitation of the potential form of Darcy's law. Expressed in terms of the driving force per unit mass, E, the flow equation is

$$\begin{aligned}\mathbf{q} &= (k\rho/\mu)E \\ &= (k\rho/\mu)[\mathbf{g} - (1/\rho)\operatorname{grad} p]\end{aligned} \tag{13a}$$

[Note that the driving force is identical to that derived from the momentum balance leading to the N-S equation, Eq. (9a).]

There exists a potential gh such that $E = -\nabla gh$ if and only if curl $E = 0$. This condition is satisfied when

$$\nabla p \times \nabla(1/\rho) = 0$$

Similarly, the driving force per unit of volume, $\mathbf{H}$, leads to the expression

$$\begin{aligned}\mathbf{q} &= (k/\mu)\mathbf{H} \\ &= (k/\mu)(\rho g - \operatorname{grad} p)\end{aligned} \tag{13b}$$

and the criterion for the existence of a volumetric force potential ρgh becomes

$$\operatorname{curl} \mathbf{H} = 0 \qquad \text{or} \qquad \nabla\rho \times \mathbf{g} = 0$$

The existence of a potential, and hence the validity of Eqs. (12), is seen to depend on rather rigid limitations on the density. For constant density, Eqs. (12) are valid and interchangeable. When there exists a unique, single-valued relation between p and ρ, Eq. (12a) is valid, but not (12b). These distinctions may become of great practical significance in cases involving, for example, the intrusion of salt water into fresh water aquifers.

There exists a voluminous literature describing the limitations of Darcy's law. This subject will not be treated here in any detail. Darcy's law may be considered as resulting from a balance between the driving forces (pressure and gravity) and the viscous resisting forces. When the inertial forces are negligible compared to the viscous forces, as is generally the case at the velocities encountered in engineering problems of flow of fluids through soils or rocks, Darcy's law can be expected to hold. At high velocities, it cannot be expected to apply. At very low gradients, deviations from Darcy's law have also been observed, but these deviations have negligible quantitative effects upon the gross flow rates.

C. The Flow Equations for Saturated Media

As stated at the beginning of this article, we need to combine a continuity equation, an equation of motion, and an equation of state to define the system.

To obtain an equation of continuity, we again consider a mass balance. The difference, however, between the situation leading to Eq. (3) and the present case lies in the fact that we now consider the total element of volume including the solid phase, rather than just the fluid phase. Thus, designating the fraction of the total volume not occupied by the solid phase by f, the equivalent to Eq. (3) becomes in vector notation

$$f\,\partial\rho/\partial t + \nabla \cdot \rho\mathbf{q} = 0 \tag{14}$$

In place of the momentum balance, we now use Darcy's law in the form

$$\mathbf{q} = -(k\rho g/\mu)\nabla h \tag{15}$$

Combining Eqs. (14) and (15) results in

$$\nabla \cdot [\rho(k\rho g/\mu)\,\nabla h] = f\,\partial\rho/\partial t \tag{16}$$

It remains to specify an equation of state. For incompressible liquids, $\rho = \rho_0$ and Eq. (16) reduces to

$$\nabla \cdot (k/\mu)\,\nabla h = 0$$

If, in addition, the soil is homogenous and the system isothermal, k and μ will be constants, so that

$$\nabla^2 h = 0, \qquad \nabla^2 p = 0 \tag{17}$$

Thus for saturated, homogeneous, and incompressible systems, Darcy's law leads to the Laplace equation, either in terms of the hydraulic head or in terms of the pressure. For compressible liquids, an expression like $p = \beta \ln(\rho/\rho_0)$ may often be used. For gases, one may set $p^m = \rho/\rho_0$ [*9*]. We note that these relations satisfy the conditions for the existence of a potential per unit weight or mass, but in general not of a potential per unit volume.

Thus we have obtained a general equation for the flow of fluids in saturated media that is considerably simpler and more useful than the Navier-Stokes equation. For incompressible liquids, the equation reduces to a time-invariant form. This does not mean that the equation cannot be applied to transient flow problems, but suggests that in such cases the time dependence must be accounted for through the boundary conditions.

D. Unsaturated Flow

The application of Darcy's law has been successfully extended to flow through unsaturated media. Again the flux is taken to be proportional to a total potential gradient, but the proportionality constant is now dependent on the degree of saturation as well as on the medium. There are valid reasons to question the validity of this hypothesis. It is likely that the dependence of flux on each component of the total driving force varies differently with, say, temperature [*10, 11*]. However, for the purpose of the present review, we shall assume that the flux is proportional to the total driving force.

It can be shown by thermodynamic reasoning that the total potential of the soil water is the sum of components due to gravity, external pressure, temperature, chemical composition, and degree of saturation. For the purposes of this article, we shall assume conditions of constant temperature, atmospheric pressure, and composition. The total potential may then be expressed as the sum of the *gravitational potential* ϕ_g and a *pressure* (or *capillary*) *potential* ϕ_c, where ϕ_c includes the effects of surface tension and adsorptive surface forces. The term pressure potential is used because ϕ_c corresponds in magnitude to the pressure in a water manometer in equilibrium with the soil water at the point considered. However, it should not be inferred that an actual pressure equal to ϕ_c can be found in the soil water. In general, the pressure potential in unsaturated soils is negative.

Expressing potential in terms of energy per unit mass, Darcy's law for unsaturated soils becomes

$$\mathbf{q} = -(K/g)\,\nabla(\phi_g + \phi_c)$$

Following the classical paper by Richards [*12*], and introducing the symbol θ for volumetric moisture content, the continuity equation may be written

$$\nabla \cdot \mathbf{q} = -\partial\theta/\partial t$$

which, when combined with Darcy's law yields

$$\nabla \cdot [(K/g)\,\nabla(\phi_g + \phi_c)] = \partial\theta/\partial t$$

The gravitational force is simply $\mathbf{g}$, so that

$$\partial K/\partial z + \nabla \cdot (K/g)\,\nabla\phi_c = \partial\theta/\partial t \tag{18}$$

Except for hysteresis, there exists a unique relation between the pressure potential and the moisture content. This enables one to write

$$\partial K/\partial z + \nabla \cdot (K/g)\,\nabla\phi_c = (\partial\theta/\partial\phi_c)(\partial\phi_c/\partial t) \tag{19}$$

Equation (19) describes the flow of water through an unsaturated soil in terms of the distribution of pressure potential ϕ_c and of two soil properties, the *unsaturated conductivity* $K(\theta)$ and the *capillary capacity* $A = \partial\theta/\partial\phi_c$. The conductivity was written as $K(\theta)$ to emphasize that it varies with the moisture content. The capillary capacity is the slope of the moisture characteristic which relates the moisture content to the pressure potential.

It is sometimes desirable to express the flow equation in terms of the moisture content rather than the pressure potential. Rewriting Eq. (18) as [*13*]

$$\partial K/\partial z + \nabla \cdot (K/g)(\partial\phi_c/\partial\theta)\,\nabla\theta = \partial\theta/\partial t$$

and defining a new soil property $D = (K/g)(\partial\phi_c/\partial\theta)$, the result is

$$\nabla \cdot D\,\nabla\theta + \partial K/\partial z = \partial\theta/\partial t \tag{20}$$

Equation (20) describes the flow in terms of the moisture content distribution and the two soil properties D and K. It is clear that Eqs. (19) and (20) are physically equivalent, and that $K = gAD$.

In case of strictly horizontal flow, Eq. (20) reduces to a form identical to Fick's law of gaseous diffusion. For this reason, D is known as the *diffusivity*, although the process described in Eq. (20) is not related to or even analogous to molecular diffusion. The dimensions of D apparently are $[L/T^2]$.

One advantage of Eq. (20) over Eq. (19) is that it frequently is somewhat easier to solve, either analytically or numerically. Another advantage is that D varies less drastically than does K, so that more confidence can be placed on solutions based on estimates of D than when estimates of K must be used. A serious disadvantage is that D tends to infinity as the soil reaches saturation, so that flow systems involving both a saturated region and an unsaturated region cannot be treated with Eq. (20), while Eq. (19) is still applicable.

It is important to reiterate that, for the solution of transient unsaturated flow problems, one requires two soil properties, say $K(\theta)$ and $D(\theta)$, which are both variables depending on θ. For steady-state unsaturated flow problems, one variable soil property suffices. For saturated flow problems, one non-varying soil property is needed.

III. Steady, Saturated Flow toward Drains

The development of theory to describe the flow of water to drains has proceeded along a variety of widely divergent paths. Some scientists have developed mathematically exact solutions to problems that have severe geometric or physical restraints. Others have used rather coarse assumptions to arrive at useful answers to practical field situations. Again, there are those who have chosen a middle path between these two.

A chronological account would suffer from lack of logical continuity; a complete accounting, from excessive length. We have elected to choose from the literature pertinent examples to illustrate some of the strengths and limitations of the analytical tools that are available, starting with the most readily tractable geometrical configurations and leading towards more difficult geometries and less exact solutions.

A. Drainage of Ponded Water

Consider the drainage from a fully saturated, homogeneous soil of great depth into a horizontal drain at depth d below the surface. As indicated in

Fig. 5, this problem requires the solution of the Laplace equation, Eq. (17), subject to the boundary conditions

$$\phi = d + t \qquad \text{for} \quad y = d \tag{21a}$$
$$\phi = s \qquad \text{for} \quad (x^2 + y^2)^{1/2} = r_d$$

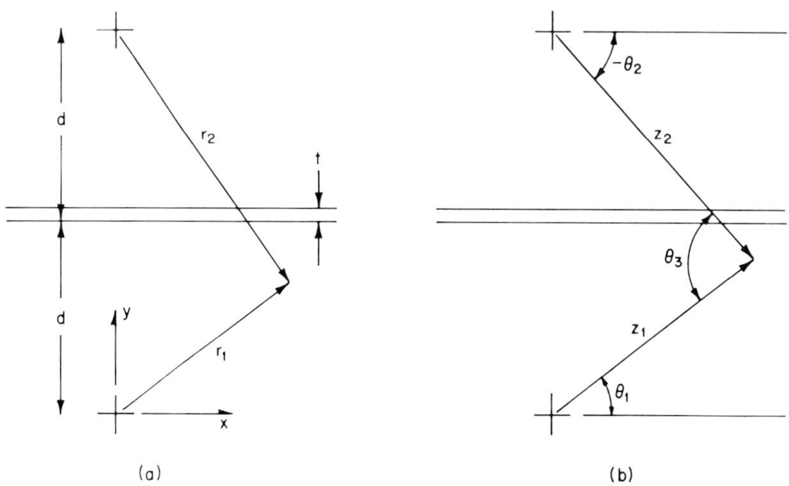

FIG. 5. Definition of symbols for flow of ponded water to a line sink. (a) in cartesian coordinates and (b) in complex notation.

where r_d represents the drain radius. Here, as through most of this article, we use the symbol ϕ to represent *potential* per unit weight, or *hydraulic head.* It has been found more tractable, however, to consider a line sink at the origin and to specify, rather than the potential along the drain circumference, only the potential at one point. Then, after a solution has been obtained, the equipotential through this point is considered as the drain circumference. Thus the second boundary condition is replaced by

$$\phi = s \qquad \text{for} \quad x = 0, \quad y = r_d \tag{21b}$$

Recognizing that, for radial flow towards a line sink, the solution of Laplace's equation is

$$\phi = A \ln r + B$$

and making use of the principle of superposition, we imagine a source at the point (0, 2d) of equal strength as the sink at (0, 0). The solution for the line source-sink combination then is seen to be

$$\phi = A(\ln r_1 - \ln r_2) + B \tag{22}$$

where A and B are constants of integration. From condition (21a), we find that $B = t + d$. Condition (21b) yields

$$A = (d + t - s)/\ln[(2d - r_d)/r_d] \tag{23}$$

so that the solution becomes

$$\phi = \frac{d + t - s}{\ln[(2d - r_d)/r_d]} \ln(r_1/r_2) + d + t \tag{24}$$

Equation (24) represents the potential distribution for a line source-link system. It may be interpreted as the potential distribution for a drain at potential s with circumference passing through the point $(0, r_d)$ when there is maintained a thin layer of water of thickness t on the surface.

To obtain the flux, or the drain discharge rate, we apply Darcy's law across the soil surface:

$$Q = \int_{-\infty}^{\infty} K\, \partial\phi/\partial y \bigg|_{y=d} dx = 2\pi K A \tag{25}$$

where A is defined by Eq. (23).

It is readily shown that the surfaces $\phi =$ constant represent a family of circles. Thus the above solution applies to a circular drain passing through the point $(0, r_d)$. Its diameter, however, is not $2r_d$, but slightly larger. Had we chosen, instead of Eq. (21b), the condition $\phi = s$ for $x = 0$, $y = -r_d$, we would have arrived at a very similar solution representing a slightly smaller drain diameter. Comparison of the two cases indicates that a compromise, in error by less than 0.1% for most conditions, is obtained in the form

$$\phi = \frac{d + t - s}{\ln(2d/r_d)} \ln\left(\frac{r_1}{r_2}\right) + d + t \tag{24a}$$

The above analysis has but limited value in itself, in that it applies only to a trivial class of drainage problems. It provides a logical building block, however, for somewhat more important problems.

The first extension requires the introduction of complex variable notation. Defining a complex potential

$$\omega = \phi + i\psi$$

the potential for radial flow becomes

$$\omega = A \ln z + B + iC$$

Here z represents the complex space coordinate $z = x + iy$. Assuming for the same problem just discussed, a solution of the form (Fig. 5b)

$$\omega = A(\ln z_1 - \ln z_2) + B + iC$$

we find by separating real and imaginary parts

$$\phi = A \ln(r_1/r_2) + B \tag{26a}$$

$$\psi = A(\theta_1 - \theta_2) + C = A\theta_3 + C \tag{26b}$$

Use of the boundary conditions yields the same values for A and B as before. We now have obtained, however, an additional equation in ψ. The lines ψ = constant represent the stream lines and ψ is called the *stream function*. Since we are interested in the amount of flow between sets of streamlines, or in the differences between ψ values rather than absolute values, the integration constant C has no significance and can be dropped.

To obtain the flux between two streamlines, ψ_1 and ψ_2 (Fig. 6), we may

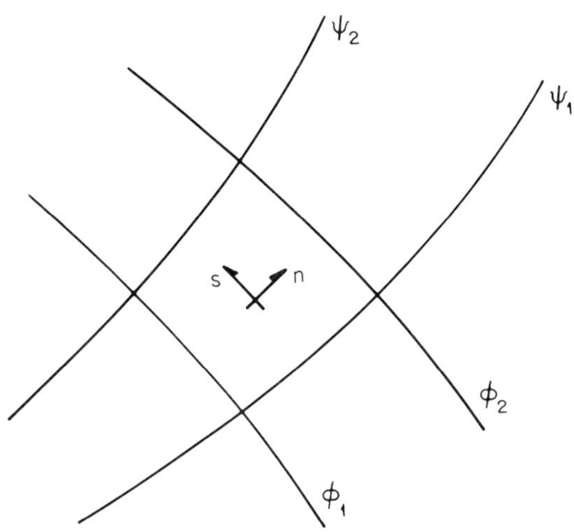

FIG. 6. Notation for determining flux between two streamlines.

write Darcy's law in differential form and integrate along an equipotential from one streamline to the next:

$$q = \int_{\psi_1}^{\psi_2} K\, \partial\phi/\partial n \Big|_{\phi=\phi_1} ds$$

Applying the Cauchy-Riemann condition, $\partial\phi/\partial n = \partial\psi/\partial s$, there results

$$q = \int_{\psi_1}^{\psi_2} K\, \partial\psi/\partial s\, ds = K(\psi_2 - \psi_1) \tag{27}$$

To obtain the total flux into the drain, we consider the values of θ_1 and θ_2 along the soil surface. At $x = \infty$, $\theta_1 = \theta_2 = 0$; at $x = -\infty$, $\theta_1 = \pi$ and

$\theta_2 = -\pi$. Thus from Eq. (26b), $\psi = 0$ at $x = \infty$ and $\psi = 2\pi A$ at $x = -\infty$. Equation (27) then yields

$$Q = 2\pi KA \tag{25}$$

The use of complex variables has resulted in identical results for ϕ and for Q as before, as it should. In addition, it has provided an equation for the streamlines which could not have been obtained readily from the earlier treatment.

A situation of greater practical interest than the single drain in a semi-infinite homogeneous medium consists of a series of parallel drains in a homogeneous soil of finite depth. Kirkham [*14*] has shown how the foregoing solution can be extended to this more general problem by the application of the method of images. This method makes use of the observation that the midplane between a line sink and a line source of equal strengths is an equipotential and that the midplane between adjacent line sources or sinks of equal strengths is a stream surface. In its simplest form, the method was used in writing Eq. (22) to conform to the boundary condition Eq. (21a). In the same manner, it can be applied to take account of an impervious layer at depth h below the surface.

With reference to Fig. 7, the image at $y = 2d$ of opposite sign to the drain at $y = 0$ forces the plane $y = d$ to be an equipotential. Similarly, an image of equal sign at $y = -2(h - d)$ would make the plane $y = -(h - d)$ a streamline, as required for an impervious layer. However, the two images together destroy the symmetry and hence their combined effect results in neither plane satisfying the required condition. This difficulty is circumvented as follows. The doublet at $y = 0$, $y = 2d$ (designated by $n = 0$) is reflected with equal sign as $n = -1$ to maintain the plane impervious layer. To reinstate a plane soil surface, the doublet $n = -1$ is reflected with opposite sign as $n = +1$, distorting the impervious layer in the process. To again reinstate the plane impervious layer, $n = +1$ is reflected with equal sign as $n = -2$, etc. The result is an infinite array of images and the proper potential distribution is found as the infinite sum of the potentials due to each image. Using the complex potential and identifying the upper image of each doublet with a prime, there results

$$\omega = A\left(\sum_{\text{even}} \ln(z_n/z_n') - \sum_{\text{odd}} \ln(z_n/z_n')\right) + B \tag{28}$$

where the summation is over $-\infty < n < \infty$.

Equation (28) represents the potential distribution for a single drain in a flooded soil with an impervious layer. To extend the solution to a series of

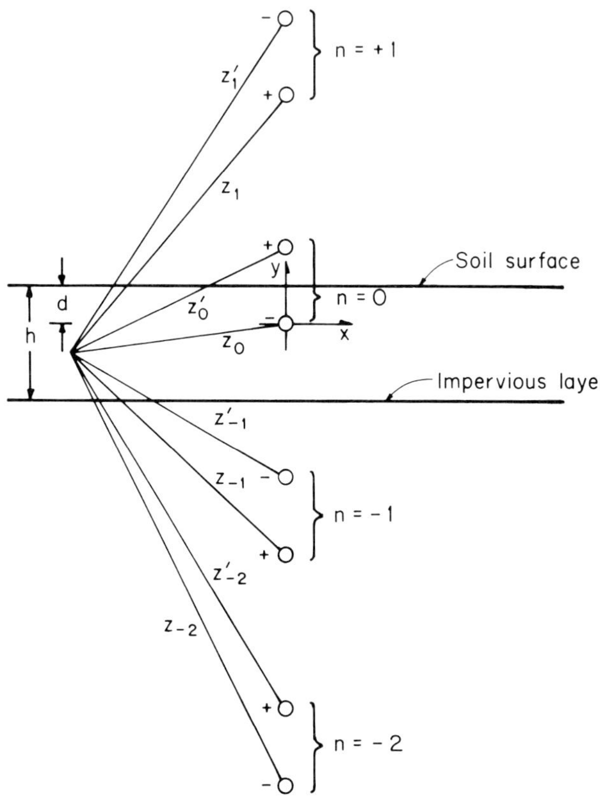

FIG. 7. Imaging to account for impervious layer.

parallel drains, it is necessary to repeat the process of imaging in the horizontal direction, resulting in a double summation.

The details of the manipulation necessary to reduce the solution to useful form and the final result can be found in the original references [*14–16*]. Suffice it, here, to state that Kirkham's results were in the form of infinite series involving hyperbolic and trigonometric functions. Vedernikov [*17*] and Gustaffson [*18*] solved the same problem, apparently independently. They expressed their results in elliptic functions.

The same method can be applied to study the flow of water into drains when the lower boundary of the flow region is a gravel artesian aquifer. The only change in the analysis is the change in the sign of the images. This problem of combined artesian and surface flow has also been treated by Kirkham [*19*].

Whereas a steady-state flooded soil drainage case is not likely to be encountered in practice, the above solutions have considerable value in several respects. Comparison with sand tank model studies [*19*, *20*] showed excellent agreement between experiment and theory; thus confidence is gained not only in the particular theoretical result being tested, but more importantly, in the validity of the underlying theoretical assumptions.

Calculations based on the theory show that a very large proportion of the flow enters over a small area centered over the drains. For example, for 6-in.-diameter drains with centers 4.5 ft deep and spaced 100 ft apart, with an impervious layer at 6 ft, the flux across the surface immediately above the drains is 140,000 times that at the midpoint. This indicates that surface grading to concentrate the water near the drains should have an important bearing on the drainage rate. It also indicates that leaching by tile drains will be effective only near the drains unless provisions are made to force the water to enter the soil at distances away from the drain.

The theory also shows that the diameter of the drain has but a minor effect on the rate of water entry. At 4-ft depth, a fourfold increase in diameter affects the flow rate by only 20%.

The theory indicates that the most important single geometric factor in drainage system performance is the drain depth. Drainage rate is roughly proportional to depth of drain until the drain approaches an impervious layer.

Thus the theory enables the designer to apply informed qualitative judgment to real-life problems, even if the field situation differs drastically from the theoretical model.

Additional results for drainage of ponded water include an analysis of the effect of imbedding the drain in an impervious layer [*21*] and the case of drainage into ditches penetrating to an impervious layer [*22*] or in a soil overlying a gravel aquifer [*23*].

Another problem in saturated flow that has been discussed in the literature concerns the effect of the crack width between clay or cement drain sections, or the size and number of the perforations in continuous drain tubes, on the inflow rate. Kirkham [*24*] presented a complicated mathematical analysis of this problem, Ernst [*25*] a simpler treatment. Whereas these analyses are of academic interest, field observations have failed to verify the theory. One explanation is that the theory assumes that the hydraulic conductivity of the soil remains constant in the vicinity of the drains, while in fact the soil frequently changes properties drastically. Flodquist [*26*] presents evidence of increasing permeability in heavy soils; MacKenzie [*27*] reports severe impedance of drainage by the formation of iron deposits near and in the drain. Thus effects other than those associated with the pure geometry of the drain appear to be of overriding importance.

B. Drainage of Steady Precipitation

The flow into drains of water that originates from a steady rate of precipitation is both more difficult to analyze and of greater interest than the ponded water case.

"Exact" solutions to steady precipitation problems have been presented by Van Deemter [*28*], Engelund [*29*], and Kirkham [*30*]. These solutions are exact in the sense that an exact mathematical solution is found to the differential equations and boundary conditions assumed applicable. In all three cases, the flow region considered is that below the water table, defined as the locus of points at atmospheric pressure. It is assumed that the soil below the water table is homogeneous and saturated and that the flux crosses the water table according to the relation $\Psi = Rx$, with R a constant precipitation rate.[3] Thus a partially saturated flow system has been reduced, by artificial restraints, to a fully saturated flow system. Since the position of the water table is not known *a priori*, even the simplified problem is difficult to solve.

1. *The Hodograph Method*[3]

A most elegant approach to the solution of two-dimensional flow problems is available in the hodograph method. It apparently was first applied to a drainage problem by Vedernikov [*31*]. Later Van Deemter [*28*] and Engelund [*29*] applied it to drains in a semi-infinite medium, and List [*32*] used it to find an approximate solution for steady drainage in the presence of an impervious boundary.

The *hodograph method* is a specialized application of the theory of conformal mapping which utilizes the theory of analytic functions. Let $w = u + iv$ and $z = x + iy$ be related as $w = w(z)$. Then $w(z)$ is said to be analytic in a region R_1 if w is single-valued everywhere in R_1 and the derivative dw/dz exists (i.e. is univalued) for every value of z in R_1. The necessary and sufficient conditions for $w(z)$ to be analytic are that the derivatives of u and v be continuous and that the Cauchy–Riemann conditions hold:

$$\partial u/\partial x = \partial v/\partial y; \qquad \partial v/\partial x = -\partial u/\partial y$$

If a function $w(z)$ is analytic in some region R_1 of the z plane, then there exists a region R_2 in the w plane such that there is a one-to-one correspondence between every point in R_1 and R_2, and we can map R_1 into R_2.

---

[3] In the discussion of the hodograph method, the force potential ϕ has been replaced by the velocity potential Φ, defined as $\Phi = K\phi$. Correspondingly, we have $\Omega = \Phi + i\Psi$. It is noted (see [4]) that a velocity potential exists only for K constant in an isotropic flow region.

The solution of steady-state flow problems is accomplished by solving a differential equation with appropriate boundary conditions. It is well known that any one set of boundary conditions will yield a unique solution. The value of conformal mapping is found in the opportunity to transform the geometry of the flow region into a region of a simpler geometry along the boundaries of which the values of the function in question are known.

The problem may be restated as follows. We search for a function $\Omega = \Omega(z)$, which expresses the *complex potential* Ω as a function of the position, z, within the flow region. If, by conformal transformations, we can identify the flow region with a simpler region in which the potential distribution is known, we have, in principle, solved our problem.

It has been found advantageous to introduce the *complex velocity*, $w = u + iv$, which like Ω, is analytic. It is defined as

$$d\Omega/dz = -w$$

But by differentiation of $\Omega = \Phi + i\Psi$ with respect to $z = x + iy$, it is found that

$$d\Omega/dz = -v_x + iv_y$$

where v_x and v_y are the velocity components in the x and y directions. Hence $u = v_x$ and $v = -v_y$. The word hodograph refers to a velocity plot. It is plain that the hodograph plane is obtained from the w plane by reflection in the real axis.

As will be shown, the boundaries of the flow region in the w plane are frequently more easily found than those in the z plane. To find the solution $\Omega = \Omega(z)$ for the given boundary conditions, it therefore may become convenient to introduce an intermediate w plane to establish the relation $w = w(\Omega)$. Having found this relation, the solution $\Omega = \Omega(z)$ is then obtained from the definition of w by integration:

$$z = -\int (d\Omega/w) + C$$

The actual details of solutions by the hodograph method are generally rather complicated and lengthy. However, the basic advantage of the method can be illustrated as follows. In the case of drainage of steady precipitation, the saturated flow region is bounded along one side by the water table, the position and shape of which is *a priori* unknown. In fact, a major part of the solution involves precisely the determination of the position of the water table. In the hodograph plane, this boundary is reduced to a circle.

We consider, after Van Deemter [*28*], the case illustrated in Fig. 8, representing parallel drains at spacing L in a homogeneous soil of conductivity K,

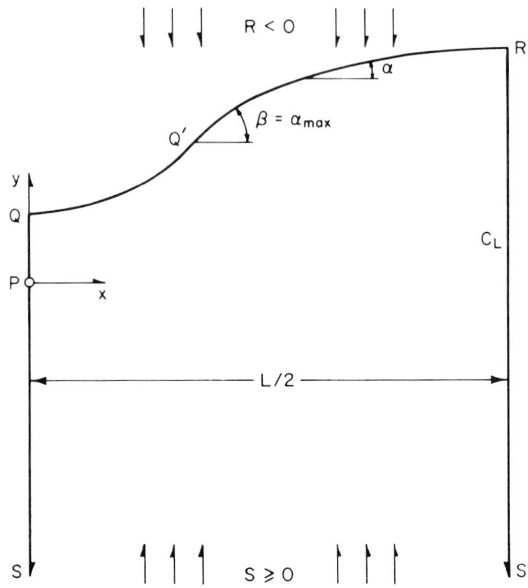

FIG. 8. Drainage of steady precipitation in homogeneous soil—the hodograph method.

and subjected to steady precipitation $R < 0$ and steady upward seepage $S > 0$ from great depth.

The boundaries PQ, RS, and SP all represent streamlines along which $\Psi =$ constant and $\partial\Phi/\partial n = 0$. In fact, along PQ we may set $\Psi = 0$; along RS, $\Psi = RL/2$; and along SP, $\Psi = (R - S)L/2$. Since these boundaries are all parallel to the vertical axis, the conditions in the w plane must include $v_x = u = 0$. At S, it was specified that $v_y = -v = S$. At P, which represents a point sink into which flows finite flux, one must have $w = \infty$.

Along the water table QR, the conditions in the z plane are $\Psi = Rx$ and $\Phi = Ky$. In the w plane, we proceed as follows. Differentiation of $\Phi = Ky$ along the water table yields

$$(\partial\Phi/\partial x)(dx/ds) + (\partial\Phi/\partial y)(dy/ds) = K(dy/ds)$$

or, with α the angle between the water table and the horizontal,

$$-u \cos \alpha + v \sin \alpha = K \sin \alpha \tag{29}$$

Similarly, from differentiation of $\Psi = Rx$ and application of the Cauchy-Riemann conditions,

$$-v \cos \alpha - u \sin \alpha = R \cos \alpha \tag{30}$$

Combining Eqs. (29) and (30) and eliminating α, there results

$$u/(K - v) = (R + v)/u$$

which may be written as

$$u^2 + [v - (K - R)/2]^2 = (K + R)^2/4$$

This last equation shows that the water table, of unknown shape or position in the z plane, transforms into a circle in the w plane.

The flow region $PQRSP$ may now be drawn in the w plane, as represented in Fig. 9. The area to the left of the shaded lines corresponds to the area inside the region $PQRSP$ in the z plane.

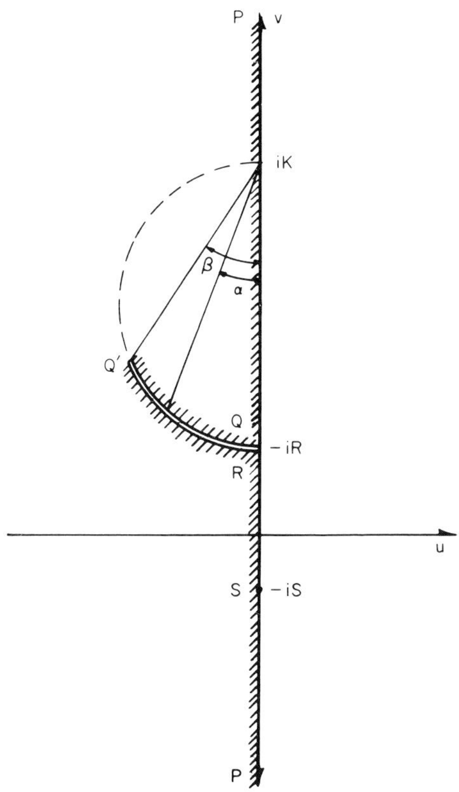

FIG. 9. Hodograph plane corresponding to flow region of Fig. 8.

The further details of the solution, obtained by transforming the w plane into a half-plane, transforming the Ω plane into another half-plane and matching the two planes, can be found in the original reference, or, in English, in van Schilfgaarde *et al.* [*33*]. Here, suffice it to state that the solution takes the form of parametric equations for Ω and z in terms of a dummy variable, the constants K, R, S, L, and the variable β representing the maximum inclination of the water table.

That this latter variable is still in the solution relates to the fact that one boundary condition still is not completely satisfied. In an actual problem, the drain will not be a line sink but will have physical dimensions. One may associate any equipotential surface around P with the drain circumference. Specifying the potential on this drain boundary (hence the pressure in the drain) will provide the additional restraint needed to fully define the problem.

To illustrate the significance of this last boundary condition, we restrict ourselves to the important special case where $S = 0$ and refer to the work of Engelund [*29*]. Along the line segment QP of Fig. 8, the pressure first increases from atmospheric pressure at Q and then decreases gradually, finally approaching $-\infty$ at P (Fig. 10). The hydraulic head is also indicated in

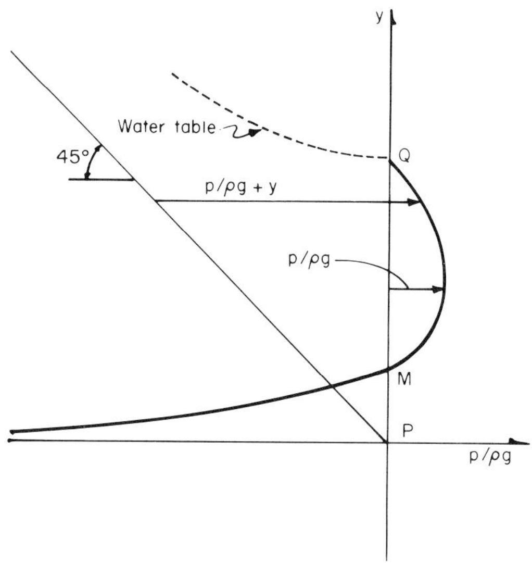

FIG. 10. Pressure distribution in plane segment immediately above line sink.

Fig. 10. If the flow system is restricted to saturated conditions, the pressure everywhere must be positive (greater than atmospheric) and the smallest drain that satisfies this condition coincides with the equipotential that passes through M. Any equipotential that crosses PQ between Q and M may be chosen as a drain circumference. Each will result in exactly the same flow pattern and water table as long as the hydraulic head in the drain is set equal to that indicated in Fig. 10 for the highest point along the drain circumference.

Restricting the discussion to the case of a just full-flowing drain, one observes that normally there will be two possible drain sizes: one passing through Q and one passing through M. The smaller of these will have a water table standing above the drain, whereas for the larger the water table will be tangent to the top of the drain. Numerical calculations show that, in practice, the smaller drain will seldom be encountered. This is in accord with the observation that water is seldom found standing above a drain. Calculations also show that changing the size of a full-flowing drain will have but a minor effect on the height of the water table midway between drains; this result supports the custom of selecting the drain size on the basis of its capacity to discharge water at the required rate, rather than on the basis of its effect on the resistance to water entry into the drain.

The drains discussed above are noncircular unlined channels that conform in shape with the equipotentials. In practice, one deals with circular tile segments or perforated tubes. The deviations from circularity result in trivial distortions if proper precautions are taken, but the effect of resistance to flow resulting from the drain channel liners cannot be so lightly dismissed. The effect of this additional resistance to flow is equivalent to an increased pressure in the drain and can thus be treated qualitatively. Unfortunately, as mentioned before, quantitative evaluation of this effect has as yet not resulted in useful results.

One frequently encounters drains that run less than full. In effect, one then deals with flow into an open channel or a ditch. To the extent that flow above the water table and surfaces of seepage can be ignored, the above hodograph solution is still completely applicable if equipotentials with a larger "diameter" than those passing through the point Q are taken as the ditch circumference. They would represent "horseshoe-shaped" ditches. Solutions for other shapes of ditches are possible. Van Deemter [*28*] treated two special cases.

When the upward seepage S is not zero, there will be a streamline that divides the flow region due to precipitation from that due to seepage. If this dividing streamline were replaced by an impervious membrane, the flow due to precipitation would not be affected. Van Deemter [*28*] made use of this observation to get an indication of the effect of an impervious layer. He found that the water table height at the midpoint for a fixed ratio K/R was not greatly affected until the average depth of the dividing streamline below the drain was less than approximately one-fourth the drain spacing.

List [*32*] used a similar device to study the effect of an impervious layer in more detail. He introduced an image drain below the actual drain, so that the actual drain would remove the precipitation and the image drain an equal flux from great depth. The dividing streamline between the two drains could then be considered as an impervious boundary. This boundary was not exactly a horizontal, but unless the drain was very near the boundary, the deviations from a plane surface were negligible. In this manner, List managed to find a solution for the flow to a series of drains above an impervious layer that exactly satisfied the boundary conditions at the water table and introduced a minor approximation along the impervious boundary, a region where the flow is sluggish and where minor changes in geometry have but insignificant effects.

2. *Exact Infinite Series Solutions*

Recently, Kirkham [*30*] has presented an "exact" solution for drainage of uniform precipitation by parallel drain tubes in soil over an impervious barrier at finite depth. Kirkham and Powers [*34*] solved the equivalent problem for open ditches.

The method used considered a partly fictitious rectangular flow region bounded by the impervious layer, the vertical planes of symmetry through the drain axis and the midplane between drains, and the horizontal plane through the highest point of the water table at the midplane. It was recognized by inspection that a solution to the Laplace equation for the potential could be written in the form

$$K\phi = \sum_{p=1}^{\infty} RaA_p \frac{\cosh(p\pi x/h)}{\sinh(p\pi a/h)} \cos(p\pi y/h)$$

$$- \sum_{n=1}^{N} RaB_n \frac{\cosh(n\pi y/a)}{\sinh(n\pi h/a)} \cos(n\pi x/a) + CRa$$

where the origin was taken on the impervious boundary midway between drains; h referred to the maximum height of the water table; a and R to the half-spacing and the precipitation rate; and A_p, B_n and C were constants to be evaluated. A corresponding equation was written for the stream function.

The unique aspect of Kirkham's solution is that, rather than specifying the value of ϕ and ψ along the upper boundary of the assumed flow region, he specified the conditions along the water table, the location of which was unknown, as $\psi = Rx$, $\phi = y$. Using this condition, together with the obvious ones along the remaining boundaries, he managed to evaluate the coefficients A_p by means of Fourier series theory and to show how the coefficients B_n could be obtained by an iterative process.

It turns out that, to obtain numerical results from this solution, one is forced to carry out elaborate and extensive calculations that are best left to a computer. Such computations apparently have not yet been made, but should make possible an evaluation of the consequences of the various simplifying assumptions that are made in other simpler and more readily usable solutions.

An earlier series solution by Kirkham [*35*], especially when combined with a graphical presentation by Toksöz and Kirkham [*36*], is more readily applied, and hence more useful, than his "exact" solution. It introduced a more drastic physical simplification, but again resulted in a mathematically exact solution of the model considered.

The model consisted of a rectangular flow region bounded by an impervious layer, the vertical planes of symmetry through the drain axis and through the midplane between drains, and a horizontal plane here designated P, through the lowest point on the water table. Here we shall restrict the discussion to the special case where this last boundary plane P passes through the drain axis.

Mathematically, the condition along the upper boundary was specified as $\Psi = Rx$. Physically, it was assumed that (1) the flow was vertically downward from the soil surface to the boundary plane P, and (2) that the head loss through this region was negligible. One may visualize such a system as consisting of a series of frictionless, rigid, vertical membranes extending from the plane P to the soil surface, with a medium between the membranes of infinite conductivity.

The form of the solution was noted by inspection in terms of a Fourier series, and the coefficients were evaluated to satisfy the boundary conditions. The result provided equations for the potential and the stream function for different values of x and y. When applied to the midpoint between drains on the plane P, the result for ϕ reduced to

$$m = (RL/K)F(d/L, r/L) \tag{31}$$

where $F(d/L, r/L)$ represented an infinite series dependent on d/L and r/L. Here d refers to the depth of the impervious layer below the drain axes, r and L to the drain radius and spacing, and m to the height of the water table above the drain axes at the midpoint.

Replacing the medium of infinite conductivity between the frictionless membranes by soil results in additional head loss and hence in a higher value for ϕ along the upper boundary. The correction turns out to be $(1 - R/K)^{-1}$, so that

$$m_u = \frac{RL/K}{1 - R/K} F(d/L, r/L) \tag{32}$$

The latter equation may be interpreted as an upper limit for the water table height—hence the subscript u. The presence of the impervious membranes forces the water to follow a longer path than it normally would and hence results in a calculated water table height higher than expected in reality.

Toksöz and Kirkham [*36*] rearranged Eq. (32) in the form

$$(m/d)(K/R - 1) = (L/d)F(d/r, L/d)$$

and presented extensive tables of $F(d/r, L/d)$ as well as a set of curves relating $(m/d)(K/R - 1)$ to (L/d) for a range of values of d/r. In this form, Kirkham's solution becomes an important tool for the solution of practical problems. Figure 11 is a graphic presentation of the F values calculated by Toksöz and Kirkham.

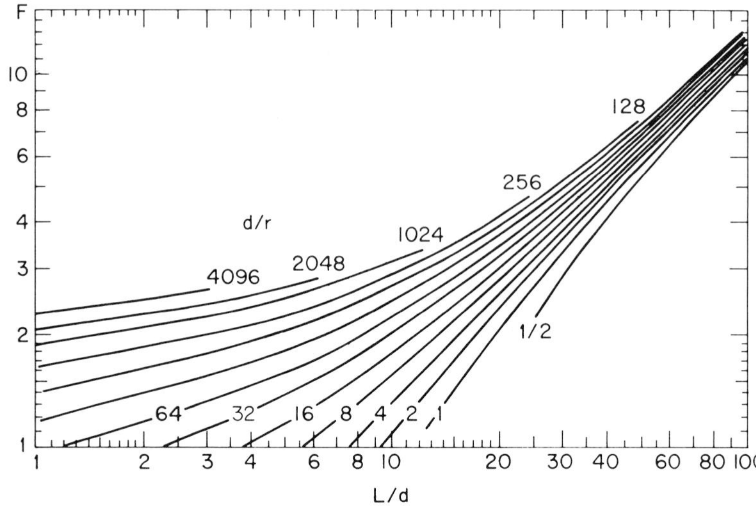

FIG. 11. Kirkham's F as function of d/r and L/d.

In the special case that the depth to the impervious layer becomes extremely large ($d = \infty$), Kirkham's solution reduces to

$$m = (LR/\pi K) \ln(L/\pi r) \tag{33}$$

In Eqs. (31–33), the drain radius r is defined as the half-width of the closed, noncircular equipotential surface at the height of the plane P. This equipotential has a value equivalent to a drain that runs half full.

Wesseling [*37*] has extended the application of the graphical Toksöz and Kirkham solution to the case of a layered soil where the conductivity K_1 of the layer above the drain axes is less than that of the soil below the axes, K_2. He showed that this only required a relabeling of the axis $(m/d)(K/R - 1)$ as $(m/d)(K_2/R - K_2/K_1)$. (This result differs from that given by Wesseling. The difference is attributed to an apparent misprint in the original article.)

Wesseling [*38*] also showed, using Kirkham's solution described here together with his more general result where the plane P passes above the drain axis, that the use of continually submerged deep drains will result in a lower water table level, given the same spacing and precipitation rate, than drains that are half full. The comparison is for equal hydraulic head in the drains. Thus even where the discharge head is controlled, say by the water level in the outlet ditch, a significant advantage is obtained in terms of either a lower water table or else a wider spacing, from deeper placement of the drains, even if this results in continuous submergence. That this effect is indeed of importance is indicated by the following numerical example.

Consider a soil with $K = 0.2$ m/day, an impervious layer at 3 m below the surface and 4-cm-diameter drains. Let it be required to maintain the water table 50 cm below the surface with a discharge rate of 7 mm/day and assume that the outlet maintains a free water surface at 100 cm below the surface. Wesseling's calculations show that this would require a drain spacing of 10 m for a drain depth of 1.0 m, but a spacing of 18 m for a drain depth of 1.5 m.

It needs to be emphasized again that the "exact" solutions discussed in the foregoing are exact only in the mathematical sense. Physically, they represent a well-defined flow system, but one that deviates from that found in practice. The most critical limitation of both the hodograph solutions and the Fourier series solution of Kirkham concerns the region of flow above the water table. The hodograph solution assumes vertical percolation through the upper soil layers into the water table at a rate equal to the precipitation rate. Thus it ignores the physical reality of one continuous flow region that includes both the partially saturated zone above the water table and the saturated zone below. Forcing the flow in this upper region vertically downward adds an unnatural restraint and hence results in an upper bound for the position of the water table. Kirkham's requirement of a uniform flux across the water table for his "exact" solution, whereas it avoids an explicit assumption as to the nature of the flow region in the unsaturated zone, has the same ultimate effect. His membrane model makes the assumption explicit and somewhat more binding by the choice of boundary plane through the lowest point of the water table.

The special value then of the mathematically exact solutions is found in the check they provide to evaluate the consequences of the assumptions made in approximate solutions. From an engineering viewpoint, the generally simpler approximate solutions are frequently preferred as long as their limitations are clearly understood.

3. *Dupuit-Forchheimer Flow*

According to DeWiest [*39*], the assumptions that lead to the so-called *Dupuit–Forchheimer theory* of groundwater were first used by Dupuit in 1857. Many others have used the same assumptions since, sometimes with reference to Dupuit and sometimes without. Forchheimer, in his Hydraulik [*40*] as well as in his earlier writing, was among those. The designation Dupuit-Forchheimer theory is probably due to Muskat [*9*]. Polubarinova-Kochina [*41*, Chapter 10] refers to it as the *hydraulic theory*.

The basic assumptions of Dupuit were that all streamlines in a system of gravity flow toward a shallow sink are horizontal and that the velocity along these streamlines is proportional to the slope of the free-water surface but independent of the depth. In addition, it is implied that the free-water surface is the streamline bounding the flow region and, in cases of an open ditch or a dam, that the water table terminates at the water level in the ditch.

It should be clear that, for the assumptions to be reasonable, their application should be restricted to flow systems that are shallow relative to their areal extent and that exhibit relatively small curvature of the free surface.

Consider (Fig. 12) a prism of base $\Delta x\ \Delta y$ and height z, taken from a flow

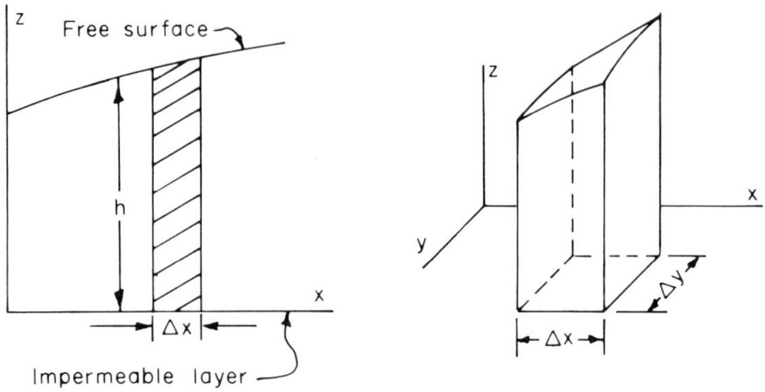

FIG. 12. Groundwater flow system in dynamic equilibrium. The shaded section at left is shown at right in three dimensions.

system bounded on top by the free surface and below by an impervious layer. It follows from the D-F assumptions that the hydraulic head, $h(x, y)$, referred to the impervious layer, coincides with the free surface $z(x, y)$. The net inflow into this prism in the x direction must be

$$K(h\ \partial h/\partial x)_{x+\Delta x}\ \Delta y - K(h\ \partial h/\partial x)_x\ \Delta y$$

and in the y direction,

$$K(h\ \partial h/\partial y)_{y+\Delta y}\ \Delta x - K(h\ \partial h/\partial y)_y\ \Delta x$$

The total net flux out of the prism must equal the precipitation (or evaporation) rate into the top surface, $R\ \Delta x\ \Delta y$. Hence addition of the two expressions above yields, in the limit as Δx and Δy vanish,

$$K \frac{\partial}{\partial x}(h\ \partial h/\partial x) + K \frac{\partial}{\partial y}(h\ \partial h/\partial y) + R = 0 \qquad (34)$$

Where there is no precipitation or evaporation ($R = 0$), the equation reduces to

$$(\partial^2 h^2/\partial x^2) + (\partial^2 h^2/\partial y^2) = 0$$

These are the differential equations resulting from the D-F assumptions.

We now apply the foregoing equation to find the shape of the water table between two parallel open ditches for constant rainfall R when the water levels in the ditches are H_1 and H_2 (Fig. 13). Equation (34) reduces to its one-dimensional form,

$$K \frac{d}{dx} (h\, dh/dx) + R = 0$$

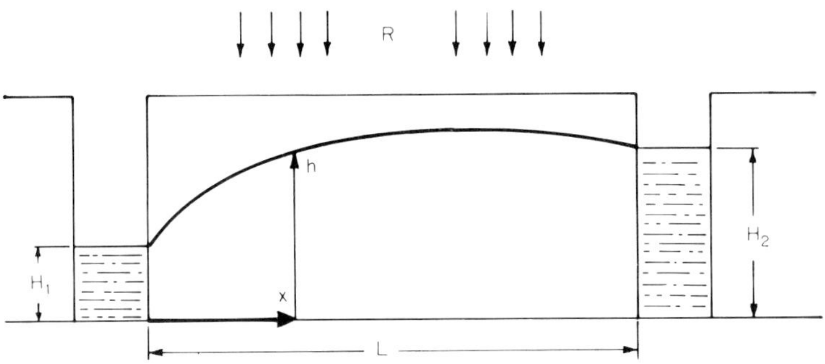

FIG. 13. Drainage to parallel ditches with unequal water levels.

which, by double integration, yields

$$Kh^2/2 + Rx^2/2 = Ax + B$$

Using the values for h at $x = 0$ and $x = L$ to evaluate the constants A and B and rearranging gives

$$K(h^2 - H_1{}^2) = [RL + K(H_2{}^2 - H_1{}^2)/L]x - Rx^2$$

For equal water levels $H_1 = H_2 = H_0$, we find

$$K(h^2 - H_0{}^2) = RLx - Rx^2 \tag{35}$$

which is the equation of an ellipse. It was first derived, albeit in a somewhat different manner, by Colding [*42*] and later, apparently independently, by a number of other workers, including Hooghoudt [*43*].

Engelund [*29*], using the hodograph technique, found an exact solution for the drainage of water into parallel horizontal drains placed at the same depth as the impervious layer. His solution

$$(x^2/K) + (h^2/R) = L^2/4K$$

is identical to that found by Colding [*42*] by the D-F theory. The spacing L here refers to the major axis of the ellipse. The spacing between the edges of the parallel drains is the focal length, or $L[(K - R)/K]^{1/2}$.

Since we are primarily concerned with the maximum height of the water table, we substitute $x = L/2$ and $h = H$ into Eq. (35) and find, upon solving for L,

$$L^2 = (4K/R)(H^2 - H_0{}^2) \tag{36}$$

In this form, the ellipse equation can be used to determine the spacing between drains to maintain a prescribed water table depth for a given precipitation rate.

Whereas the equation was derived for open ditches, it may also be applied to tile drains if the hydraulic head in the drains is substituted for the water level in the ditches.

Frequently it is reasonable to assume that the drains will not operate under pressure. This assumption, together with the introduction of the symbols $m = H - H_0$ for the rise of the water table above the drain axes and d for the height of the drain axes above the impervious base, changes the equation to

$$\begin{aligned} L^2 &= (4K/R)(H + H_0)(H - H_0) \\ &= (4K/R)(2d + m)m \\ &= 8Kdm/R + 4Km^2/R \end{aligned} \tag{37}$$

This form of the ellipse equation is frequently encountered in the literature.

The ellipse equation provides a far simpler tool for drainage design than do the more sophisticated solutions of Kirkham or Van Deemter. This simple solution can be used advantageously under circumstances where field conditions closely approximate the assumptions used in the theory. Of paramount importance is the requirement that the depth of the water-bearing stratum be small as compared to the spacing between drains, for the assumption of horizontal flow ignores the loss of head due to convergence of flow towards the drain. Inspection of Eq. (37) indicates that as the depth increases without limit, the theory results in an indefinitely increasing spacing. Obviously, this is an absurdity. Even for relatively shallow systems, however, the error introduced by the simplification can be quite significant. To avoid this difficulty, several modifications have been proposed. We shall now examine some of these modifications.

4. *Hooghoudt's Solutions*

Hooghoudt [*43, 44*] considered the general problem of steady drainage towards ditches or tile drains in a homogeneous soil underlain by an impervious layer as being intermediate between the case of radial flow when the impervious layer is at great depth, and that of horizontal flow for a very shallow impervious layer. By judiciously combining the solution for radial flow, similar to our Eq. (24), with that for horizontal flow, Eq. (37), he managed to develop a solution applicable for any arbitrary depth to the impervious layer.

Many of the details of Hooghoudt's elaborate analyses are summarized elsewhere [*16*, *33*]. He restricted his analyses to circular drains running half-full, which were considered tile drains when the diameter was small and ditches when the diameter was large. He applied the radial flow equation to the region near the drains and the ellipse equation to the region away from the drains to determine the water table drop (or head loss) in each region, and added the two amounts to find the total drop.

To separate the two regions, he chose the vertical plane along which the potential difference was a minimum according to the radial flow equation. This turned out to be the plane $x = d\sqrt{2}/2$ with the origin at the drain and d the depth below the drain of the impervious layer.

Hooghoudt obtained his solutions in terms of infinite series and presented elaborate tables to simplify their use. An interesting innovation, since widely adopted, was the introduction of an equivalent depth. The equivalent depth d_e is that value of d which, when substituted into Eq. (37), will result in the correct relationship. In other words, Hooghoudt retained the form of the ellipse equation for application over the full range of conditions $0 < d < \infty$, but provided an appropriate correction for convergence of flow by introducing the parameter d_e. He presented lengthy tables of d_e for a range of conditions.

The equivalent depth concept offers a powerful means for extending the usefulness of the D-F theory to a wider range of problems. This concept has proved particularly valuable for nonsteady flow problems for which solutions based on other than the D-F theory are difficult to obtain.

To make the equivalent depth tables more readily available, van Schilfgaarde [*45*] presented a graph of d_e as a function of d and L for a drain diameter of 5 in. More recently, Moody [*46*] gave the relation

$$\frac{d}{d_e} = 1 + \frac{d}{L}\left(\frac{8}{\pi}\ln\frac{d}{r} - \alpha\right), \qquad 0 < d/L \leq 0.3 \tag{38}$$

in which

$$\alpha = 3.55 - 1.6d/L + 2(d/L)^2$$

For $d/L > 0.3$, Moody suggested

$$L/d_e = 8\,[\ln(L/r) - 1.15]/\pi$$

The use of $\alpha = 3.4$ in Eq. (38) introduces an error in α of less than 5%. Moody's equations are based on a reexamination of Hooghoudt's series solution.

Wesseling [37] has compared the Hooghoudt solution with that of Toksöz and Kirkham [36] and found that the two never differed by more than 5%. It seems remarkable that the relatively simple model used by Hooghoudt yielded results so close to the mathematically and physically more elaborate and less approximate solution of Kirkham. Wesseling's finding certainly adds to the confidence with which the Hooghoudt solution may be applied.

5. *Other Approximations*

It has long been known [9, p. 398] that the D–F theory often results in the correct expression for flux, but does not properly predict the shape of the free surface. In the case of drainage to an open ditch, or seepage through a dam, D–F theory forces the free surface to exit at the point where the water level stands in the ditch. A surface of seepage is not accounted for. Charnii, as reported by Polubarinova-Kochina [41, p. 281], avoided the assumption of a horizontal velocity independent of depth and showed that the D–F result for flux through a dam was indeed correct. With symbols as in Fig. 14,

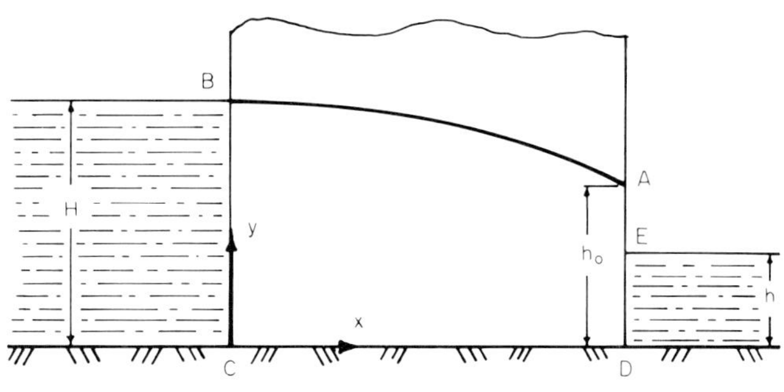

FIG. 14. Symbols used in Charnii's evaluation of flux through a dam.

Charnii established the following boundary conditions:

$$\text{along } AB: \quad \phi(x, y) = y \tag{39}$$

$$\text{along } BC: \quad \phi(0, y) = H \tag{40}$$

$$\text{along } DE: \quad \phi(L, y) = h \tag{41}$$

$$\text{along } EA: \quad \phi(L, y) = y \tag{42}$$

Designating the value of y at the free surface as $y = Y$, he then considered the flux across any plane $x = x_1$ as

$$q = -\int_0^Y K\, \partial\phi/\partial x\, dy$$

Applying the rule for differentiation under the integral sign,

$$q = -\frac{d}{dx}\int_0^{Y(x)} K\phi\, dy + K\phi(x, Y)\, dY/dx$$

$$= -\frac{d}{dx}\int_0^{Y(x)} K\phi\, dy + KY\, dY/dx$$

where we have used Eq. (39). Integration results in

$$\int_0^L q\, dx = qL = \left[-\int_0^{Y(x)} K\phi\, dy + KY^2/2\right]_{x=0}^{x=L}$$

To evaluate the right-hand side of this expression at $x = L$, we use Eqs. (41) and (42):

$$-K\int_0^h h\, dy - K\int_h^{h_0} y\, dy + Kh_0{}^2/2 = -Kh^2/2$$

At $x = 0$, the application of Eq. (40) results in

$$-\int_0^H KH\, dy + KH^2/2 = -KH^2/2$$

Hence,

$$qL = K(H^2 - h^2)/2$$

This result is identical to that obtained more simply from the D-F theory by integration of the expression

$$q = -Ky\, dy/dx$$

between the limits $(0, H)$ and (L, h). But Charnii's approach makes unnecessary the D-F assumptions and allows for the existence of a surface of seepage. Thus it establishes the validity of the simpler D-F result for flux.

The same reasoning was used by Guyon [*47*] to find a solution for drains resting on an impervious layer. He had to introduce the additional assumption, however, that the potential drop along the vertical plane midway between drains was negligible. For steady precipitation, Guyon's result is again identical to the ellipse equation, Eq. (36). To extend his solution to the more general case where the drains do not rest on an impervious layer, however, Guyon introduced Hooghoudt's equivalent depth and arrived, as expected, at the same result as Hooghoudt. Thus Guyon avoided in part the D-F assumptions that the velocity is invariant with depth and proportional to the slope of the water table, and permitted an explicit accounting for a surface of seepage.

Dagan [*48*] followed the same qualitative reasoning as did Hooghoudt. He accepted the D-F theory as sufficiently accurate for the region far from the drains, but used potential theory to derive an expression for potential drop near the drains. Specifically, Dagan obtained a solution to the potential flow problem by linearizing the water table boundary condition to

$$\partial\phi/\partial y = R/K$$

Physically, this amounts to restricting the flow region to that below the plane through the lowest point on the water table, and it reduces the flow region to a rectangular region. In this rectangular region of depth h, Dagan determined separately a line-sink confined-flow potential and a potential due to steady precipitation. The former was obtained by the method of images as illustrated in Fig. 7, except that all images had the same sign as the drain. The total potential was then obtained by superposition and its value along the upper boundary of the flow region was taken as the height of the water table.

As the horizontal distance from the drain increases, the inaccuracy introduced by the linearized boundary condition also becomes more objectionable. To overcome this difficulty, Dagan restricted the potential solution to the region $0 < x < 2h$, and in the region $2h < x < L$, he applied the ellipse equation. Combination of the two yielded the rather simple result, after some additional simplifying assumptions, with the symbols of Eq. (36),

$$H^2 - H_0{}^2 = (R/K)[L(L - 2\alpha H_0)]$$

When applied to a tile drain intersected by the water table, the symbol α in this equation is a function of the drain diameter $2r$ and the depth to the impervious layer h, or $\alpha_1 = \alpha_1(r/h)$. When the water table does not intersect the drain ($d \neq 0$ in Fig. 7), $\alpha_2 = \alpha_2(d/h)$ and an additional relation is needed to define the drain diameter, of the form

$$r/h = f(K/R, d/h, L/h)$$

When the drain is a full ditch of depth d and width $2r$, then $\alpha_3 = \alpha_3(r/h, d/h)$. Dagan has presented tables and graphs of α_1, α_2 and α_3 and for the determination of r.

Whereas Dagan's analysis is fairly complicated, application of his results is relatively simple and no doubt superior to the ellipse equation. Like Hooghoudt's solution, it overcomes the limitation of the ellipse equation due to the effect of convergence of flow towards the drains. It does involve a number of approximations, however, the consequences of which need to be determined by numerical comparisons. To my knowledge, such calculations have not been made.

A quite different approach was used by Ernst [*25*], who considered that the total resistance to flow consisted of the vertical resistance due to flow from the water table into the groundwater system, the horizontal resistance related to lateral flow, and the radial resistance associated with convergence of flow toward the drain or ditch. When the water table does not deviate greatly from a horizontal, and when the average depth of the water-bearing stratum D is less than one-fourth the spacing, the head loss between the midpoint between drains and the water level in the drains is given by

$$\Delta h/R = (D^*/K) + (L^2/8Kd) + Lw \tag{43}$$

In this equation, D^* refers to the (ill-defined) depth through which the vertical head dissipates. It may be taken as approximately equal to, but slightly larger than, the water table height m. Generally, $D^* \ll L^2/8d$, and hence may be neglected. The term $L^2/8Kd$ is derived from the D-F assumptions and accounts for the lateral resistance $L/4Kd$. The radial resistance w enters into the last term.

The radial resistance for semicircular drains may be calculated approximately as

$$w = (1/\pi K)\ln(d/\pi r)$$

and for shallow flat-bottomed ditches as

$$w = (1/\pi K)\ln(4d/\pi B)$$

where B is the ditch width.

Equation (43) was restricted to the condition $4d < L$. When d becomes very large, the radial resistance term also increases without limit. Using potential theory, Ernst derived an alternate solution for the case $d \to \infty$ in the form

$$h = (RL/\pi K)\ln(L/\pi r) \tag{44}$$

In the region $0.25 < d/L < 0.5$, use of either equation will give approximately the same answer. For $d/L > 0.5$, it is adequate to assume $d \to \infty$.

IV. Nonsteady Saturated Flow toward Drains

In nature, steady-state conditions are seldom encountered. Whereas a varying rainfall rate may sometimes be treated fruitfully as a steady state by using the mean value [*49*], many problems of practical interest must take explicit account of the transients.

The solution of nonsteady saturated flow problems requires, besides a knowledge of the saturated conductivity, some assumption as to the volume of water stored in or drained from the soil. It is common practice to consider a constant (i.e., time-independent) specific yield or drainable porosity. Childs [*49*], among others, has pointed out that this assumption can lead to significant errors, and quantitative determination of an appropriate value for porosity is wrought with difficulties. In view of the complicated nature of the problem, however, it remains the most practical assumption available for the development of theory useful in design.

Although the flow equation derived from potential theory, Eq. (17), is time-independent, it is possible to apply it to transient problems by introducing time-dependent boundary conditions. Another approach that has been used advantageously is based on the application of D-F theory. By either approach, it is possible to study the rate of fall of the water table from a known initial condition in the absence of precipitation, or its behavior as the result of an irregular, time-varying precipitation pattern.

A. Falling Water Table and D-F Theory

If the position of the water table is permitted to vary, the differential equation resulting from the D-F assumptions is obtained from Eq. (34) by replacing the right-hand side by $f\,\partial h/\partial t$, with f the drainable porosity. For two-dimensional problems, such as parallel drains, this equation reduces to

$$K\frac{\partial}{\partial x}(h\,\partial h/\partial x) + R = f\,\partial h/\partial t \tag{45}$$

Because this nonlinear differential equation is difficult to solve, it is frequently linearized [*41*, p. 500]. One method of linearization is based on the assumption that the variation in h is small compared to that in $\partial h/\partial x$, leading to

$$K\bar{h}\,\partial^2 h/\partial x^2 + R = f\,\partial h/\partial t \tag{46}$$

Here $\bar{h}$ is considered constant. A second type of linearization is obtained by writing $u = h^2$ and assuming that $\bar{h} = \sqrt{u}$ is constant:

$$\begin{aligned} K\sqrt{u}\,\partial^2 u/\partial x^2 + R &= f\,\partial u/\partial t \\ K\bar{h}\,\partial^2 u/\partial x^2 + R &= f\,\partial u/\partial t \end{aligned} \tag{47}$$

Both methods of linearization have been applied to drainage problems. Both assume that the change in depth of the water-bearing stratum is small compared to the average depth. Thus their application must be restricted to $\Delta h \ll \bar{h}$, as well as $\bar{h} \ll L$, as required by the D-F assumptions. Here Δh refers to the change in h with time as well as with horizontal position.

1. *Boussinesq's Problem*

Boussinesq [*50*] and, subsequently but independently, Glover [*51*] showed that Eq. (45) could be solved for the water table drawdown from an initial condition characterized by the cessation of a previously constant precipitation rate ($R = 0$) when the drains penetrate to the impervious layer of a shallow aquifer ($d = 0$) by assuming that the variables are separable, or (Fig. 15)

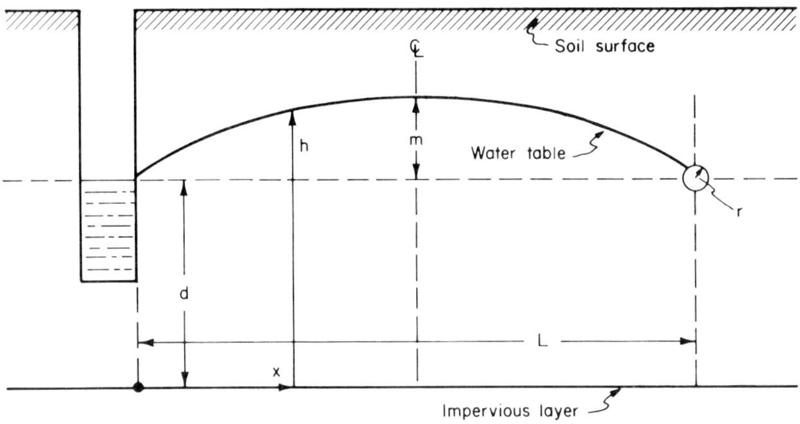

FIG. 15. Symbols used in falling water table analysis.

$$h(x, t) = X(x)T(t) \tag{48}$$

Substitution of Eq. (48) into Eq. (45) yields the two ordinary differential equations

$$(d^2X/dx^2) + [(dX/dx)^2/X] = c \qquad (dT/dt)/T^2 = fc/K$$

subject to the conditions

$$\begin{aligned} h &= h_0 && \text{at } x = L/2; \; t = 0 \\ h &= 0 && \text{at } x = 0, L; \; t \geq 0 \\ dh/dx &= 0 && \text{at } x = L/2; \; t \geq 0 \end{aligned} \tag{49}$$

In terms of the drain spacing, the solution to these equations yields

$$L^2 = (9Kt/2f)m_0 m/(m_0 - m) \tag{50}$$

where the coefficient has been rounded by less than $\frac{1}{2}\%$ for simplicity [*33*]. The initial condition implicit in this solution is a parabolic water table of maximum height h_0.

2. *Linearized D-F Solutions for Deep Aquifers*

When the drains do not penetrate to the impervious layer ($d \neq 0$), the Boussinesq type of solution breaks down because the second boundary condition of Eqs. (49) can no longer be satisfied. To circumvent this difficulty, several investigators–Glover [51], Krayenhoff van de Leur [52], and Maasland [53]–have used Eq. (46).

Starting from an initially flat water table, the boundary conditions may be written as

$$\begin{aligned} h = h_0 = d + m_0 \quad & \text{for} \quad 0 < x < L \quad \text{at} \quad t = 0 \\ h = d \quad & \text{for} \quad x = 0, \ L \quad \text{at} \quad t \geq 0 \\ dh/dx = 0 \quad & \text{for} \quad x = L/2 \quad \text{at} \quad t \geq 0 \end{aligned}$$

Using a Fourier series, the solution becomes

$$h - d = (4m_0/\pi) \sum_{n=1,3,5\ldots}^{\infty} (1/n) \sin(n\pi x/L) \exp(-n^2\pi^2 KDt/fL^2)$$

Here, following Glover, we have taken the average depth $\bar{h}$ equal to $D = d + m_0/2$, or the average initial depth of the water-bearing stratum.

For all but the smallest time periods, all terms but the first may be neglected, resulting in the simple expression, at $x = L/2$,

$$L^2 = \pi^2 KDt/f \ln(4m_0/\pi m) \tag{51}$$

Before proceeding, it is appropriate to reemphasize the limitations inherent in the D-F theory. Clearly, Eqs. (50) and (51) cannot be valid for large depths. In Eq. (50), this restriction is not serious since the depths to which drains can be dug are physically limited. In Eq. (51), however the same limitations apply as for the steady state and the same method can be used to circumvent them. The use of Hooghoudt's equivalent depth, whether in the form of Eq. (38) or by some other means, extends the usefulness of Eq. (51) to any value of d (and hence D). Furthermore, if the difference $(m_0 - m)$ is large, the use of a constant D based on m_0 becomes objectionable. This problem may be resolved, in an approximate manner, by stepwise solution of the equation for smaller increments $(m_0 - m)$, with a new value of m_0, and hence D, for each step, although this procedure violates the initial condition imposed.

Tapp and Moody [54] observed that the initial water table shape encountered in the field was more often parabolic than flat. They modified the initial condition accordingly and found a solution for spacing that differed from Eq. (51) only in that the numerical constant 4 was replaced by approximately 3.7.

Werner [55] chose the second linearization, Eq. (47), in preference to Eq. (46). In an important paper, he presented a series of solutions to a more general class of problems than that discussed here. He included as additional variables the slope of the impervious layer, unequal heights of the water levels in the ditches and a precipitation not necessarily zero ($R \neq 0$). His solutions were obtained by means of Laplace transforms and, in general, took the form of sine series. He showed how, by superposition, his solutions could be applied to problems of intermittent recharge, or a variable precipitation pattern. Maasland [*53*], using the first linearization, also derived a series of solutions for various recharge patterns. Since Eqs. (46) and (47) have mathematically exactly the same form, one set of solutions can readily be derived from the other. The problem of intermittent and variable recharge patterns will be discussed in a later section.

3. *Nonlinearized D-F Solutions For Deep Aquifers*

It was stated above that, when the drains do not rest on the impervious barrier, the Boussinesq solution of Eq. (45) breaks down. Van Schilfgaarde [*45*] found a solution for this case in the form

$$L^2 = 9A^2 \frac{K(d+m)(d+m_0)t}{2f(m_0 - m)} \tag{52}$$

where A is defined in terms of an incomplete beta function. With negligible error, A can be represented by the equation

$$A^2 = \frac{m_0(2d + m_0)}{(d + m_0)^2} \tag{53}$$

This solution, however, still fails to satisfy the second of the boundary conditions, Eq. (49). This failure causes negligible error for small time increments, but results in greatly overestimated drawdown for larger time increments. Fortunately, this shortcoming may be circumvented [*56*] by considering the drawdown process as a sequence of small steps. Solving Eq. (52) for t and substituting the expression for A from Eq. (53), there results

$$t = \frac{2fL^2}{9K} \frac{(m_0 - m)(d + m_0)}{m_0(2d + m_0)(d + m)}$$

This equation may be interpreted as representing the time increment Δt required for the drawdown increment $\Delta m = m_0 - m$. The next time increment may then be obtained from the next drawdown increment Δm, where now m_0 is assigned to the water table height at the end of the first increment. The total drawdown is the summation of the successive drawdown increments. In the limit, as Δm becomes infinitesimally small, this summation reduces to the integral

$$t = \frac{2fL^2}{9K} \int_m^{m_0} \frac{dm}{m(2d + m)}$$

$$= \frac{fL^2}{9Kd} \ln \frac{m_0(2d + m)}{m(2d + m_0)} \tag{54}$$

Equation (54) represents a valid solution to the D-F differential equation, Eq. (45), and satisfies the boundary conditions, Eq. (49). If used in conjunction with the equivalent depth concept discussed earlier, it provides a practical solution to a complicated problem in simple form.

Brooks [57] attacked essentially the same problem in a different manner. He adopted as the initial condition a flat water table, rather than the parabolic shape implied by Eq. (54), and applied the *PLK* method of successive approximations utilizing a perturbation parameter. His results could not be presented in closed form, but took the form of a family of curves of the relative drawdown as a function of time for different depths to the impervious layer.

B. Falling Water Table and Potential Theory

Several investigators have searched for a solution to the falling water table problem using potential theory. The most successful of these attempts seems to be one by Kirkham [58]. He initially assumed the same physical model as that which led to Eq. (31); that is, the region above the plane through the drain axes was considered as one of infinite vertical conductivity and zero horizontal conductivity. Thus the initial condition was taken as the equilibrium condition under steady precipitation,

$$h(x) = (RL/K)F(x)$$

where now $F(x)$ is still a function of x. In Eq. (31), this function was evaluated at $x = L/2$. Kirkham then "froze" the shape of the streamlines and considered the discharge through each rigid streamtube as the precipitation was stopped. Replacing the medium of infinite conductivity with soil would raise the water table (or hydraulic head referred to the plane through the drain axes) to a value $b(x) > h(x)$. Writing for the discharge through any streamtube of top width Δx the three expressions

$$\begin{aligned} q &= -f(db/dt)\,\Delta x \\ &= [K(b-h)/b]\,\Delta x \\ &= (hK/LF)\,\Delta x \end{aligned}$$

one can readily derive the differential equation

$$db/dt = (-K/f)/(1 + LF/b)$$

Integration yields

$$b/b_0 = \exp[(b_0 - b - Kt/f)/LF] \tag{55}$$

When the head loss through the layer above the drains may be ignored, so that $h = b$, Eq. (55) reduces to

$$h/h_0 = \exp(-Kt/fLF) \tag{56}$$

Using the tabulated values for $F = F(L/2)$ (see [*34*] or Fig. 11), either Eq. (55) or (56) may be used to calculate the rate of drawdown at the midpoint between drains. The shape of the water table as a function of time can be determined similarly by evaluating $F(x)$. Equation (55) gives an upper limit for the water table height at any time, because its derivation was based on the rigid configuration of the flow tubes. In reality, these flow tubes would change configuration in such a manner as to reduce the head loss. Equation (55) cannot be solved explicitly, whereas Eq. (56) is readily solved. The difference between the two results generally will be so minor that Eq. (56) should be preferred. It may be considered as nearly an upper limit.

Dagan [*59*] and Guyon [*47*] also used potential theory for the study of falling or fluctuating water tables. Dagan considered parallel drains in homogeneous soil of infinite depth and, as in his steady-state solution (46), linearized the boundary condition imposed on the free surface. It now became

$$f\,\partial\phi/\partial t + K\,\partial\phi/\partial y = -R$$

His solution involved a Volterra integral of the first kind, which was solved by a numerical technique. Thus his results were presented in dimensionless tables and graphs. Guyon used the same assumptions described in the discussion of steady-state problems. His result was essentially the same as Eq. (54).

An entirely different approach was taken by Bouwer and van Schilfgaarde [*60*] who proposed an extremely simple, albeit theoretically inelegant, concept for adaptation of any steady-state solution to a falling water table. Their method is amenable to steady-state equations or to graphical or tabular solutions obtained from theory, analogs, models, or field experiments.

The basic assumption is that the flux per unit surface area in the steady precipitation case can be equated to the instantaneous flux in the falling water table case for the same water table height, or

$$f\,dm/dt = -R$$

This relationship implies a uniform flux distribution which, in turn, requires that the water table drop at the same rate at all points. To account for a change in water table shape, or a rate of drop varying with distance from the drain, a coefficient C is introduced. Given a relationship between R and m for steady-state conditions, the above equation can be integrated to determine the time dependence of m.

For example, Eq. (37) may be rewritten

$$R = (8Kd_e m + 4Km^2)/L^2 \qquad (37)$$

and this expression for R substituted in

$$Cf\,dm/dt = -R \tag{57}$$

Integration yields

$$\frac{Kt}{f} = \frac{CL^2}{8d_e} \ln \frac{m_0(m + 2d_e)}{m(m_0 + 2d_e)} \tag{58}$$

Similarly, Eq. (31) yields, after integration,

$$Kt/f = CLF \ln(m_0/m) \tag{59}$$

To apply the procedure to $R - m$ relations that are available numerically or graphically rather than in terms of a mathematical function, we rewrite Eq. (57) as

$$\Delta t = Cf\,\Delta m/\bar{R} \tag{60}$$

and determine, from an $R - m$ plot, the mean value of R corresponding to each increment Δm. Time increments are then readily computed and aggregated for successive increments Δm.

To determine the spacing between drains from a prescribed drawdown rate, one can use a family of $R - m$ curves for different spacings. An example is shown in Fig. 16 prepared from nomographs based on a set of numerical iteration solutions [*60*]. Suppose we desire to determine the spacing required to drop the water table from $m_0 = 6$ to $m = 5$ ft in 0.8 day, given $K = 7.5$ ft/day, $f = 0.15$ and $d = 5$ ft. The lines $m_0/d = 1.2$ and $m/d = 1.0$ are plotted on the graph; the quantity $\bar{R}$ is calculated from Eq. (60)—using $C = 0.8$—and plotted as $\bar{R}/K = 0.02$. The required spacing is read from the graph as $L/d = 25$.

The factor C may be considered as the ratio of the average flux between drains to the flux midway between drains. As illustrated schematically in Fig. 17, this ratio varies from values considerably greater than unity shortly after an initially ponded condition, to unity for intermediate conditions, to about 0.8 for small values of m ($0.02 < m/L < 0.08$). Unless circumstances warrant a more specific choice, experience has shown that a value of $C = 0.9$ will yield good results.

Kirkham's "upper limit" solution, Eq. (56), is identical to Eq. (59) with $C = 1.0$. This is entirely consistent with the above discussion which would suggest $0.8 < C < 1.0$, since Kirkham's initial condition was a steady-state profile. Also, Eq. (58) is identical to Eq. (54) with $C = 8/9$, which again is consistent. These comparisons show that the simplified model leads to results that are mathematically and physically compatible with more complicated analyses. The simplicity of its application and the wide flexibility afforded through its adaptability to many sources of data for steady-state $R - m$ relations, make this simplified model attractive to theoreticians and practitioners alike.

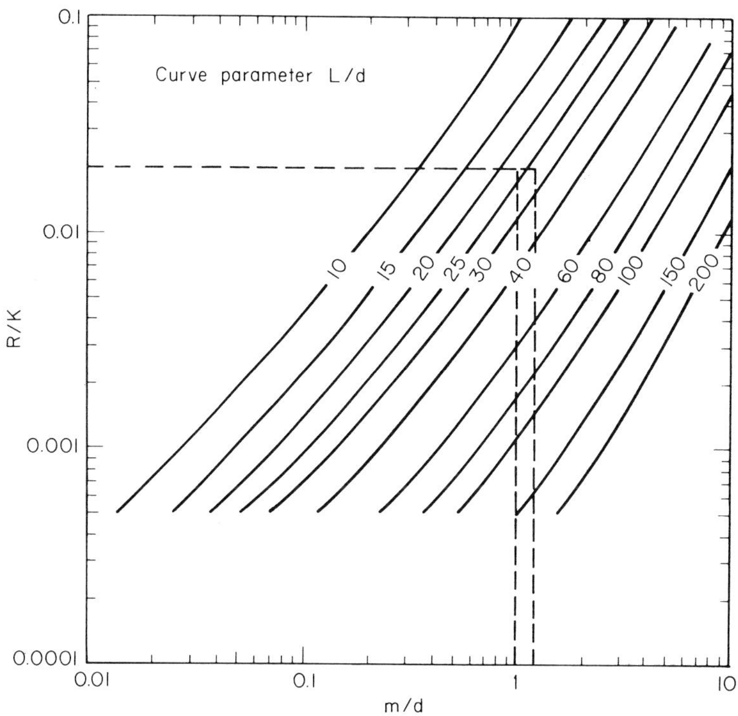

FIG. 16. Relation between R/K and m/d for evaluation of desired drain spacing.

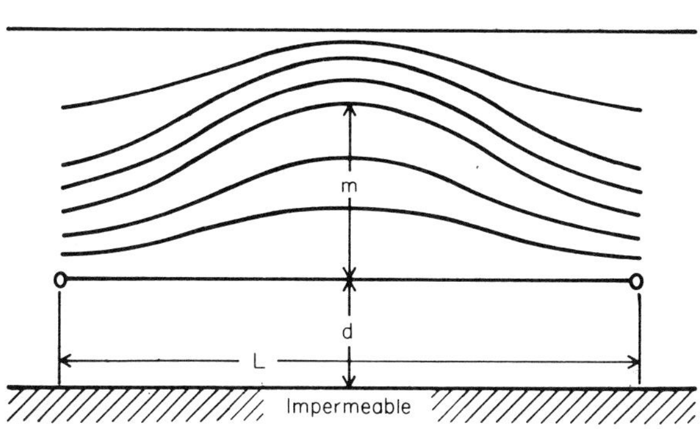

FIG. 17. Change in shape of water table with time.

C. Intermittent Recharge

Having considered the fall of the water table after precipitation ends, we shall now consider its behavior in response to an intermittent recharge pattern. This is the situation encountered when a drainage system provides relief from precipitation occurring in an irregular pattern and also applies to the recharge to the water table from periodic irrigations.

For the moment, we shall consider recharge to the water table as synonymous with precipitation. Later, we shall take account of storage of water in the soil above the water table and losses by evaporation. It is assumed, however, that all gravity drainage takes place through the drainage system, without allowing for deep seepage losses.

Werner [*55*] presented a solution to this problem that, in principle, was based on superposition of solutions obtained from the linearized D-F equation, Eq. (47). Maasland [*53*] illustrated a similar application for a regular square-step irrigation or precipitation pattern, based on Eq. (46). Krayenhoff van de Leur [*52*], who studied the response to an arbitrary precipitation pattern approximated by a bar histogram, recognized a system parameter, which he called the *reservoir coefficient*, that characterizes the drainage-soil system; we will introduce an equivalent concept. A different approach was used by De Zeeuw and Hellinga [*61*]. They proposed the use of a standard precipitation pattern, chosen to represent the desired frequency of occurrence, to determine the required design parameters for a drainage system.

Here we restrict the discussion to the analysis by van Schilfgaarde [*62*]. Combining Eqs. (31) and (57) yields the differential equation

$$-fC\, dm/dt = mK/LF(d/L,\, r/L)$$

Introducing the symbol $B = fCLF/K$, this equation may be written

$$dm/dt + m/B = 0 \tag{61}$$

The factor B, with dimensions of time, combines the geometric restraints of the system, as fixed by L and F, with the soil properties f and K into one system-defining constant. It is equivalent to the reservoir coefficient of Krayenhoff van de Leur who showed how either drain discharge or water table height data could be used to determine B from the slope of the recession curve.

Rather than integrating Eq. (61) as was done to obtain Eq. (59), a forcing function is imposed that represents an impulse. Making use of the unit step function $S(t - t_1)$, defined as $S = 0$ for $t < t_1$ and $S = 1$ for $t \geq t_1$, Eq. (61) is modified to

$$dm/dt + m/B = (m_0/\varepsilon)[S(t - t_1) - S(t - t_1 - \varepsilon)]$$

This introduces a pulse of magnitude m_0 applied over a time interval ε beginning at time t_1. Assigning the initial condition $m(0) = 0$ and letting ε approach zero, there results after integration

$$m(t) = m_0 \exp[-(t - t_1)/B]S(t - t_1) \tag{62}$$

This equation, identical to Eq. (59), may now be interpreted as the solution for an impulse m_0 at $t = t_1$. To treat a continuous precipitation rate $r(t)$ over the time interval $t_0 < t < b$, we divide the interval into n equal subdivisions, and recognize that the pulse height m_0 may be considered as caused by a constant precipitation rate R over the time increment ε, so that $m_0 = R\varepsilon/f$. As n increases indefinitely, an equation like Eq. (62) may be written for each increment. By superposition, this leads to the equation

$$\begin{aligned} m(t) &= \lim_{n\to\infty} \sum_{k=1}^{n} \exp[-(t - t_k)/B]R(t_k)\,\Delta t/f \\ &= (1/f)\int_{t_0}^{b} r(\tau)\exp[-(t-\tau)/B]\,d\tau, \qquad t \geq b \end{aligned} \tag{63}$$

where t_k refers to the beginning of the kth time increment and τ is used to differentiate between the variable of integration τ (representing time) and the time $t \geq b$ at which the water table height is sought.

Equation (63) gives a general solution for the water table height resulting from a continuous precipitation pattern, $r(t)$. Whether the equation can be integrated depends, of course, on the form of $r(t)$. In the special case that $r(t) = R$, a constant, over the period $t_0 < t < b$,

$$m(t) = (RB/f)\{\exp[-(t - b)/B] - \exp[-(t - t_0)/B]\}, \qquad t \geq b \tag{64}$$

The course of the water table during and following a storm of any pattern can readily be determined, either from Eq. (63) by stepwise integration or from Eq. (64) by superposition, if the storm pattern can be reduced to a bar histogram. To apply the theory to the design of drainage systems, however, requires several additional considerations.

First, a distinction must be made between precipitation and recharge to the water table. One way to determine the amount of recharge is to calculate the amount of precipitation that exceeds the amount of evapotranspiration from the surface and the available soil moisture storage capacity [*63*]. Assuming that no surface runoff occurs, this excess represents the water table recharge.

Next, in order to simplify the calculations, it is convenient to use a uniform time increment for each of the bars of the histogram. Then if R_n represents the net accretion rate during the nth time period, the height of the water table at the end of the Nth time period is given by

$$m_N = (B/f)[\exp(1/B) - 1]\sum_{n=1}^{N} R_n \exp[-(N - n + 1)/B] \tag{65}$$

With Eq. (65), one can determine the water table height at the end of each of an arbitrarily long sequence of time increments. Choosing a convenient interval such as a day, climatological data (precipitation and temperature suffice) can be used to follow the daily water table movement over a period of years. Thus frequency distributions can be developed of water table heights associated with a range of design parameters, and these in turn make possible the selection of a design based on a prescribed risk. An example of the type of data that can be developed is given in Fig. 18.

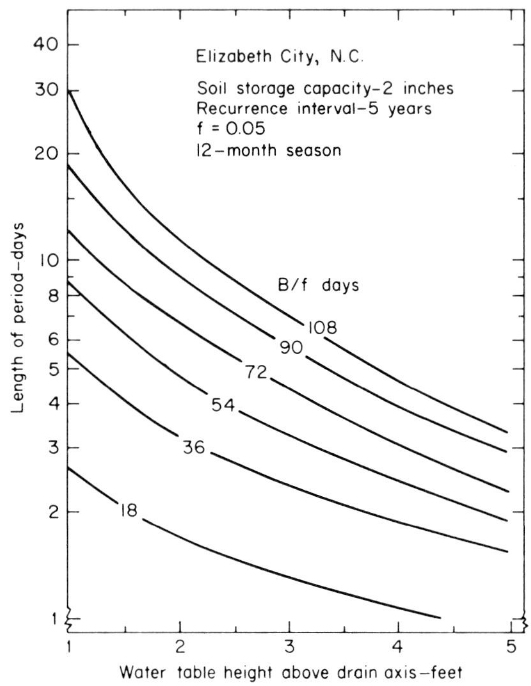

FIG. 18. Lengths of periods given water table heights are exceeded once in five years for a range of B/f values. Evapotranspiration calculated for $f = 0.05$ and a soil moisture storage capacity of 2 in.

V. Partially Saturated Flow

The separation of flow in the unsaturated region from saturated flow is an arbitrary and philosophically unsatisfying device. Physically, one has to deal with one continuous flow system with continuous flow lines and a continuous potential distribution. The only justification for making a distinction is found in the analytical complexity associated with unsaturated flow.

The appropriate differential equation for a system that includes a region of unsaturated flow is

$$\partial K/\partial z + \nabla \cdot (K/g)\, \nabla \phi_c = \partial \theta/\partial t \qquad (18)$$

A solution to this equation requires that a unique relationship exist between K and θ. If hysteresis can be ignored, as in the case of drainage from an initially ponded and saturated condition, such a unique relation may frequently be established. In general, it will take the form of a graphic or tabular representation, which will preclude an analytical solution. For such cases, a number of numerical solutions have been obtained. Examples include the digital computer solutions of Sewell and van Schilfgaarde [*64*] and the electric analog work of Bouwer [*65*]. These numerical techniques will not be discussed in this article.

Gardner [*66*] found that for a number of soils the conductivity could be expressed as a function of the pressure $p = \rho\phi_c$ by a relation of the form

$$K = a/[(-p)^n + b]$$

The value of n seems to vary from a minimum of about 1.5 for fine-textured soils to about 8 for uniform sands. Evidently, the ratio a/b is the saturated conductivity. This relation, for some values of n, has enabled the analytical solution of Eq. (18) for the one-dimensional flow in vertical columns. Application to two-dimensional drainage problems, however, has so far been restricted to finite difference methods on digital computers.

Brooks and Corey [*67*] found that the relative conductivity (expressed as a fraction of the saturated) could be described by a pore size distribution index and the bubbling pressure. The pore size distribution index may be taken as the negative slope of a double-logarithmic plot of relative conductivity versus pressure, and it corresponds approximately with the exponent n of Gardner. The bubbling pressure is a measure of the maximum pore size forming a continuous network of flow channels in the medium. It may be compared to Bouwer's critical pressure, described later. The treatment by Brooks and Corey is more general than that of Gardner and has the additional advantage, by application of similitude, of lending itself to meaningful model experiments wherein a limited number of materials can be used to represent a range of widely different soils.

To account for unsaturated flow in a meaningful way in relation to drainage design, however, still requires an approximate treatment rather than an exact solution of Eq. (18). The simplest concept that has been proposed [*16, 33*] makes use of the capillary fringe. In a column of soil in equilibrium with a water table, the soil will be nearly saturated in the region just above the water table, and the degree of saturation will then decrease rapidly with elevation and finally approach again a nearly constant value. The concept of a capillary fringe presumes that the transition zone is sufficiently sharp that one can define a height below which the soil is virtually saturated but at a pressure less than atmospheric and above which the saturation is low enough to severely restrict flow. The presence of a capillary fringe of height w does not alter the total hydraulic head to be dissipitated, but it enlarges the flow region by a height w. In the case of D-F solutions, this results in a simple adjustment of the results already presented. For example, Eq. (35) would become

$$K[(h + w)^2 - (H_0 + w)^2] = RLx - Rx^2$$

and Eq. (36)

$$L^2 = (4K/R)[(H + w)^2 - (H_0 + w)^2]$$

Childs [*68*, *69*] made a similar modification to the Van Deemter hodograph solution of steady precipitation. As expected, Childs' expression for the potential differed from Van Deemter's only by the constant w.

A somewhat different and physically more satisfying approximation was proposed by Myers and van Bavel [*70*]. Restricting themselves to predominantly horizontal flow, they defined an effective thickness h_e of a saturated layer with the same transmissibility as the entire region above the water table as follows:

$$h_e = (1/K_s) \int_{z_0}^{z_1} K \, dz$$

Here K_s represents the saturated conductivity, K the unsaturated conductivity, z_0 the elevation of the water table and z_1 that of the soil surface. The advantages of h_e over w are that it is quantitatively well defined whereas w is not, and that it is based on the soil's ability to transmit water rather than on the degree of saturation.

The application of the h_e concept is, of course, the same as described for w. A third term that has been used is *critical pressure*. Bouwer, who first defined critical pressure in qualitative terms, later [*71*] defined it analogously to h_e as

$$p_c = (1/K_s) \int_0^{p_w} K \, dp$$

where p_w is the (negative) soil water pressure at the soil surface. From data selected from the literature, Bouwer calculated the critical pressure for 28 different soils ranging from sands to clay, and found a range from -10 to -115 cm, with a mean of around -41. Eliminating one soil reduced the range to -10 to -71 cm. These values enable one to make an order of magnitude estimate of the effect of flow in the unsaturated region on the drainage rate and demonstrate that this effect is large enough to result frequently in highly significant errors if not taken into account. Bouwer proposed that, rather than adjust the upper boundary of the flow region, it may be more convenient in calculations to lower the position of the (equivalent) impervious layer by an amount equal to $-p_c$.

Gardner [72] proposed the use of a mean diffusivity for the solution of the falling water table problem. Writing Eq. (18) for one dimension in the form

$$\frac{\partial}{\partial x}(K\,\partial h/\partial x) = \partial\theta/\partial t$$

where $gh = \phi_g + \phi_c$, Gardner assumed $\partial h/\partial\theta$ approximately constant at the low pressures likely to be found in the neighborhood of a water table. This led to the form

$$\partial h/\partial t = \frac{\partial}{\partial x}(K\,\partial h/\partial\theta \cdot \partial h/\partial x)$$

Considering now the product $\bar{D} = K\,\partial h/\partial\theta$ constant rather than $\partial h/\partial\theta$, he wrote

$$\partial h/\partial t = \frac{\partial}{\partial x}(\bar{D}\,\partial h/\partial x)$$

Using as boundary conditions $h = 0$ at $x = 0$ and $\partial h/\partial x = 0$ at $x = L/2$, with initial condition $h = L/2$, a solution can be obtained in the form

$$Q_t/Q_\infty = 1 - (8/\pi^2)\exp(-\bar{D}\pi^2 t/L^2)$$

Q_t and Q_∞ represent the amounts of water drained per unit length of drain at time t and at final equilibrium, respectively.

Whereas Gardner's solution can be shown to have the same form as Glover's [see Eq. (51)], one significant difference is that here no reference is made to a water table. Only the average water content of the soil is considered. Comparisons with data from field experiments and from digital computer solutions tend to indicate that Gardner's approach gives reliable results in terms of outflow. The use of water content instead of water table height may well turn out to be preferable in terms of practical application. The drainage requirements of plants are related to aeration, and hence to percent saturation. Furthermore, the water table height can vary drastically with negligible changes in water content.

VI. Measurement of Soil Properties

The subject of soil property measurement falls outside the scope of this review. However, since the usefulness of drainage theory depends on the ability to obtain meaningful estimates of the pertinent physical soil properties, it is appropriate to make mention of a few selected measurement techniques that are particularly pertinent to drainage design. Of primary importance in this connection is the measurement of hydraulic conductivity.

In contrast to the situation encountered in foundation engineering or earthen dam design, agricultural drainage is concerned with the hydraulic conductivity of undisturbed soil. Thus laboratory methods for the determination of hydraulic conductivity of disturbed samples are likely to be meaningless. Even when undisturbed samples are brought into the laboratory, results have frequently been disappointing. The core samples represent a very small sample of the large mass of soil and they are likely to be greatly affected by relatively small scale inhomogeneities such as root channels. Also, though frequently called "undisturbed," such cores are subject to edge effects, compaction and similar disturbances. An example of the limited usefulness of hydraulic conductivity determinations from cores is found in an analysis of the results from some 10,000 cores reported by Mason *et al.* [*73*]. They found that, when five samples were taken from one site, one could place the hydraulic conductivity into one of three broad classes (high, medium, low) with a probability of 95% of being correct; into four classes with an 80% probability of having it in the proper class; and into seven classes with a 30% probability.

A number of methods have been developed for the *in situ* determination of saturated hydraulic conductivity. These have been reviewed in detail by Kirkham [*74*, *75*] and Bouwer [*76*]. Suffice it here to present the principles of some of these methods.

A. Hydraulic Conductivity Measurement in Presence of Water Table

Where field conditions are such that a relatively steady water table is found above the level at which it is desired to know the hydraulic conductivity, one can use the rate of flow into or out of a small cavity or well for its determination. Three single cavity methods have been used: the tube method, piezometer method and auger hole method.

The *auger hole method* utilizes an uncased cylindrical hole, dug to some depth below the water table. The water is baled out and then the rate of rise of the water level in the hole is measured. The diameter of the hole must be sufficiently small so that the volume of water drained does not significantly affect the location of the water table. A 4-in.-diameter hole is most common.

The *piezometer method* uses a cased hole with an unlined cylindrical cavity extending below the casing. Again, the rate of rise of the water level into the well after baling is determined. Various dimensions have been used, but most common is a 2-in.-diameter piezometer with a 4-in.-long cavity.

The *tube method* is a variation of the piezometer method without a cavity. As originally developed, an 8-in.-diameter tube was used.

For all three methods, the flow of water into the well can be expressed as

$$dQ/dt = KhS$$

where dQ/dt represents the flow rate and h the difference in elevation between the water table and the height of the water level in the well. S is a shape factor, to be discussed later.

From the geometry of the well, it is obvious that

$$dQ/dt = -\pi r^2 \, dh/dt$$

where r is the radius of the well. Elimination of Q from these two expressions and integration between the levels h_1 and h_2 occurring at times t_1 and t_2 results in

$$K = \frac{\pi r^2}{S} \frac{\ln(h_1/h_2)}{t_2 - t_1}$$

Thus a series of measurements of h at known times t enables the calculation of a series of estimates of K.

The shape factor S describes the influence of the geometry of the system on the hydraulic impedance. The total impedance to flow is the product $1/KS$. To determine numerical values of S mathematically is extremely difficult, but they are readily obtained from electric analog studies. For the tube and piezometer methods, the geometry does not change during the period of measurement, so that one value for S suffices. For the auger hole method, the flow system changes gradually as the water level in the auger hole rises, and S must be known as a function of h. Appropriate S values are available in the literature [*74*].

The auger hole method measures primarily the horizontal component of hydraulic conductivity and results in an integrated value over the depth of the hole. The tube method measures primarily the vertical component in the zone directly below the bottom of the tube. The piezometer method yields a vector combination of the horizontal and vertical components depending on the length-to-diameter ratio of the cavity. By judicious choice of geometries and methods, the vertical and horizontal components can be separated for anisotropic soils.

The field methods described must be used with care and with a clear understanding of some of their limitations. The quoted references can be consulted for restrictions on geometries and increments in water level rise, and for suggestions on equipment and techniques. Notwithstanding their limitations, field methods offer significant advantages over laboratory methods and have been used extensively both in research and in operational drainage design. An important advantage relates to the size of the sample that is tested. Taking the zone in which 80% of the head-loss occurs as the sample size, it has been shown that the relative soil volume sampled for the various methods is 1 : 35 : 100 : 270 : 1400 for a 3- by 3-in. core sample, a 1- by 4-in.

piezometer cavity, a 2- by 4-in. piezometer cavity, an 8-in. tube, and a 4- by 30 in. auger hole.

Even with the auger hole method, the soil sampled represents but a small part of the total flow medium. The sample size can be further improved by means of multiple-well methods, but the complications introduced negate their usefulness for practical design purposes. In some cases, it is possible to measure the discharge from an existing drainage system in relation to the water table height and from these determinations calculate an effective conductivity for a field. Such measurements in turn can then be used to predict the effectiveness of additional drainage works, or possibly can be extrapolated to similar soil conditions on neighboring fields.

The determination of representative values of hydraulic conductivity for use in design equations remains a serious problem where there is opportunity for major advances. Just the same, more extensive application of currently available field methods can result in significant improvement in prevalent design methodology.

B. Hydraulic Conductivity Measurement in Absence of Water Table

Frequently, the hydraulic conductivity must be determined for a soil at a time when the soil is not saturated. In the planning of new irrigation projects for arid regions, for example, one does not expect to encounter a high water table. Two methods, out of several that have been proposed, will be mentioned.

The *double-tube method* [*76*] uses two concentric tubes that are inserted into an auger hole dug to the depth at which the measurement is desired. These tubes are covered with a water-tight lid with a stand pipe for each tube. The tubes are filled with water and equal water levels are maintained at the top of both stand pipes to create a zone of positive water pressure in the soil just below the tubes. After a wetted zone has been established, two sets of measurements are taken. First, the water supply to the inner tube is shut off and the rate of fall of the water level in the stand pipe is determined. Then, after the water levels have again been brought to the top of the stand pipes, the rate of fall of the inner-tube stand pipe is again determined, but this time the water level in the outer-tube stand pipe is allowed to drop at the same rate by manipulating a bleeder valve. From the difference in rate of fall between the two measurements, the hydraulic conductivity can be calculated.

The principle upon which the method is based is that one can distinguish two separate superimposed flow systems. One consists of primarily downward flow from both cylinders into the soil. The other consists of flow from the outer annulus into the inner tube. The flux from the outer into the inner tube is determined from the difference in rate of fall between the two sets of measurement. This system has a well-defined and fixed geometry and its impedance, as in the piezometer method, depends on a shape factor and the conductivity. The shape factor has been determined by electric analog.

Experience has led to a choice of 2 for the ratio between the diameters of the concentric tubes. Diameters of 5 and 10 in. have proven satisfactory.

The double-tube method seems the most satisfactory field method available at present for measuring the hydraulic conductivity in the absence of a water table. It has, however, severe limitations. It requires relatively elaborate equipment, including up to 50 gallons of water per test, and it deals with a rather small sample volume (on the order of 10 times that of a 3- by 3-in. core sample). Since the method is based on infiltration, the soil generally will not be saturated during the test and the hydraulic conductivity obtained will accordingly be less than the saturated hydraulic conductivity. Whether this should be a matter of concern depends on how the data are to be used. In evaluating a seepage problem, the effective conductivity may well be better approximated by the double-tube technique than it would be by a measurement at saturation.

The *air-entry permeameter* was designed to obtain a measure of the air entry value of a soil. The air entry value refers to the pressure head of the soil water at which air at zero gage pressure enters soil with a continuous vapor phase. For all practical purposes, the air entry value is equivalent to the critical pressure discussed previously, and hence can be used as a "capillary fringe" height.

The permeameter [77] consists of a 10-in.-diameter cylinder pressed some 4 in. into the soil. The cylinder is capped with an air-tight lid equipped with a standpipe with reservoir, a vacuum gage, and an air escape valve. A relatively high (± 40 in.) head is used to force water into the soil after all air has been let out of the system. When the wet front has reached approximately the bottom of the cylinder, a supply valve in the standpipe is shut off. The water pressure in the cylinder will begin to decrease, reach a minimum and then increase again as air moves upward through the wetted zone and emerges at the soil surface. This minimum pressure, referred to the elevation of the wet front, is the air entry value or critical pressure.

To use the air-entry permeameter for the measurement of hydraulic conductivity, we apply Darcy's law. Prior to the closing of the supply valve, we have a one-dimensional flow system bounded above by the soil surface and below by the wet front. The potential at the soil surface is known from the height of the water level in the reservoir. The potential at the lower boundary is later calculated from the air entry value, p_a. The pressure in the water just behind the wet front is the water entry (or air exit) pressure, p_w. Bouwer has found [77] that $p_a = 2p_w$ with remarkable regularity. Thus it is only necessary to determine the flow rate—namely, the rate of fall of the water level in the reservoir—immediately prior to the closing of the supply valve to have all the information to calculate the hydraulic conductivity, K. The value of K obtained in this manner refers to the hydraulic conductivity during sorption, or at less than saturation. The saturated conductivity can be estimated as twice the K for sorption. This seemingly arbitrary relation is not based on theoretical considerations, but simply results from inspection of field and laboratory data. The combined maximum error resulting from the use of the two factors ($p_a = 2p_w$ and $K_{sat} = 2K_{sorption}$), according to Bouwer [*77*], may range from 10 to 20%.

VII. Application of Flow Theory to Drainage Design

By design, the foregoing discussion has been restricted to the theory of flow to drains. Yet our interest in this theory is based on the need for better methods to design drainage systems. To date, the vast majority of drainage systems in the United States is designed without explicit reference to the theory that has been developed. Part of the reason for this situation can be found in the fact that flow theory addresses itself only to the question of hydraulic system performance, without reference to either the requirements for drainage or the difficult problem of describing the necessary soil parameters quantitatively.

The objective of artificial drainage of agricultural land, in the words of Bouwer [*78*], is to help maximize the economic return from the farming enterprise. To translate this objective into engineering terms requires the identification of the function of the system and the establishment of numerical criteria for performance of this function.

In humid areas, the primary functions of drainage systems are two: to provide an environment favorable for plant roots and a soil-physical condition conducive to economic farming operations. The first of these may be described in terms of water table height, aeration, soil moisture content, or soil temperature, or a combination of these. The latter concerns timeliness of farm operations, trafficability and similar considerations. In irrigated regions, the primary function relates to establishing or maintaining a favorable salt balance in the soil.

Aside from a quantitative description of the functional requirements, the engineer needs to know the physical properties of the soil before he can design the system. Of primary concern are reliable estimates of hydraulic conductivity, depth of the water-conducting layer and pore space distribution.

Thus, in broad terms, drainage design requires agronomic determination of requirements, soil-physical quantification of the soil medium, engineering prediction of system performance and, finally economic evaluation of alternatives. Of these four components needed for the successful design of the total system, the analysis of the flow system is by far the easiest to evaluate with the tools at hand at present. The theory reviewed in this article, whereas far from exhausting the subject, is sufficiently well advanced to answer most questions accurately enough for practical need. Our limited knowledge of crop response to the environment, on the other hand, makes the formulation of quantitative design criteria a hazardous task, and methods for characterizing soil properties *in extenso* have not been developed to the extent that they enable routine use within the economic restraints encountered [*79*]. Economic evaluation presupposes, of course, the existence of sufficient information to successfully complete the preceding steps in the design procedure.

In view of the situation just described, we have argued that the need for further research is not so much in refinements in flow theory as in defining agronomic requirements and developing methods of characterizing soils. Another area in need of work is in the development of imaginative methods to translate established concepts in flow theory into forms that are readily adapted to design procedures.

Not mentioned so far is another important aspect of drainage system design, namely that dealing with the mechanics of drainage materials and construction techniques. This area is not discussed here because, although it is always of importance and sometimes overridingly so, it can readily be treated separately and does not enter significantly into the main theme of this article.

There are, of course, instances where rational design methods have been developed and applied successfully. A brief review may serve to suggest paths that may be explored productively towards developing techniques that will be adapted more extensively.

In climates where long-duration, low-intensity precipitation is prevalent enough to warrant the assumption of a steady state, design can be based on maintaining a fixed, time-averaging, water table depth and the disposal of a constant precipitation rate. Formulas such as Eqs. (31) and (37) provide the necessary relationship between m, R, and L. In the Netherlands, many years of observation have established that a depth to water table of 20 to 30 cm for grassland, 40 to 50 cm for cropland, or 60 to 70 cm for truck crops gives good results, and a mean precipitation rate of 7 mm/day is used. Thus the specification of two parameters enables the adaptation of a particular drainage system to the soil characteristics encountered and land use proposed. This is in contrast to the specification of just a drainage coefficient, in terms of a water removal rate, which can be applied only to the determination of a drain size with sufficient discharge capacity.

Falling water table equations such as Eq. (51) or, preferably, (54) have been applied effectively in the design of drainage systems in association with irrigation in arid regions [*80*]. The problem here may often be reduced to providing sufficient drainage to assure that the water table over time does not tend to rise above a preset level. From the irrigation practice, the number and amounts of irrigations can be determined; climatological data and cropping practices provide the information necessary to calculate the evapotranspiration. A "bookkeeping" system can then be set up to follow the position of the water table.

An additional consideration that must here be taken into account concerns the salt balance. The quantity of salt removal by the drainage water must equal the amount brought in by the irrigation water to maintain the status quo. In reclaiming saline soils, removal must exceed addition [*78*].

A method was indicated earlier [see Eq. (65) and Fig. (17)] for adapting existing flow theory to the conditions encountered most commonly in humid areas. Whether the details of this analysis as presented are followed, or alternate treatments preferred [e.g. *81*], the important point is that flow theory permits the determination of frequency distributions of water table heights. In principle, this approach enables the derivation of design criteria which lend themselves to economic evaluation. It also provides the framework for casting the questions that need answering concerning the response of crop plants to drainage. Experience may well show that the selection of the water table height as the primary parameter should be modified. There is no reason, however, why a similar analysis could not lead to criteria in terms of average water content of the root zone, or even oxygen content or potential oxygen diffusion rate. An adaptation of Gardner's work [*72*], for example, might be appropriate.

There is little reason to question the validity of the theory presented *per se*. Sufficient, although admittedly infrequent, checks of theory against field data [*47*, *58*, *82*], against model data [*81*], and of one theory against another [*45*], have been made to have confidence in the results. Where explicit solutions are not readily available, the use of modern computers makes possible the development of numerical solutions to any desired accuracy for almost any well-defined specific problem. The restraint that keeps drainage theory from being applied more widely in actual field design is not the inadequacy of the theory but the inability to find the input necessary to apply the theory.

From the viewpoint of the engineer, drainage system design concerns consideration of the total system and recognition of the interaction among its components. Progress in design techniques along rational lines depends on the explicit recognition of these interactions so that design criteria can be evolved that lend themselves to quantitative application as inputs to theoretical models.

REFERENCES

1. President's Science Advisory Committee, "The World Food Problem," Vol. II. p. 406. The White House, 1967.
2. U.S. Dept. of Agriculture, "Agricultural Statistics." 1966.
3. LAMB, H., "Hydrodynamics," 6th Ed. MacMillan, London, 1932. Reprinted, Dover, New York, 1945.
4. BASSETT, M. A., "A Treatise on Hydrodynamics." Deighton Bell, Cambridge, 1888. Reprinted, Dover, New York, 1961.
5. PRANDTL, L., and TIETJENS, O. G., "Fundamentals of Hydro- and Aeromechanics." McGraw-Hill, New York, 1934. Reprinted, Dover, New York, 1957.
6. HUBBERT, M. K., Darcy's law and the field equations of the flow of underground fluids. *J. Petrol. Technol.* **8**, 222–239 (1956).
7. SCHEIDEGGER, A. E., "The Physics of Flow Through Porous Media." Univ. of Toronto Press, 1957. Revised, MacMillan, New York, 1960.
8. LAGERWERFF, J. V., NAKAYAMA, F. S., and FRERE, M. H., Hydraulic conductivity related to porosity and swelling of soil. *Proc. Soil Sci. Soc. Am.* **33**, 3–11 (1969).
9. MUSKAT, M., "The Flow of Homogeneous Fluids Through Porous Media." McGraw-Hill, New York, 1937. Second printing, Edwards, Ann Arbor, Michigan, 1946.
10. COREY, A. T., and KEMPER, W. D., Concept of total potential in water and its limitations. *Soil Sci.* **91**, 299–302 (1961).
11. RAATS, P. A. C., and KLUTE, A., Development of equations describing transport of mass, momentum and energy in soils. UNESCO Symp., "Water in the Unsaturated Zone." Wageningen, Netherlands (1966) (in Press).
12. RICHARDS, L. A., Capillary conduction of liquids through porous media. *Physics* **1**, 318–333 (1931).
13. KLUTE, A., A numerical method for solving the flow equation in unsaturated materials. *Soil Sci* **73**, 105–115 (1952).
14. KIRKHAM, D., Flow of ponded water into drain tubes in soil underlying an impervious layer. *Trans. Am. Geophys. Un.* **30**, 369–385 (1949).
15. KIRKHAM, D., Artificial drainage of land: streamline experiments, the artesian basin: II, *Trans. Am. Geophys. Un.* **21**, 587–594 (1940).
16. LUTHIN, J. N., ed., Drainage of Agricultural Lands. Am. Soc. Agron. Mono. 7, Madison, Wisconsin, 1957.
17. VEDERNIKOV, V. V., Theory of filtration in soils and its application in problems of irrigation and drainage. In Russian. Pub. Bur. of Public Works, p. 248, Moscow, USSR, (1939).
18. GUSTAFSSON, Y., Untersuchungen über die Strömungsverhältnisse in gedräntem Boden. *Acta Agr. Suecana* **2(1)**, 1–157 (Stockholm) (1946).
19. KIRKHAM, D., Artificial drainage of land: streamline experiments, the artesian basin: III. *Trans. Am. Geophys. Un.* **26**, 393–406 (1945).
20. KIRKHAM, D., Pressure and streamline distribution in waterlogged land overlying an impervious layer. *Proc. Soil Sci. Soc. Am.* **5**, 65–68 (1941).
21. KIRKHAM, D., Reduction in seepage to soil underdrains resulting from their partial embedment in, or proximity to, an impervious substratum. *Proc. Soil Sci. Soc. Am.* **12**, 54–59 (1948).
22. KIRKHAM, D., Seepage into ditches in the case of a plane water table and an impervious substratum. *Trans. Am. Geophy. Un.* **31**, 425–430 (1950).
23. KIRKHAM, D., Seepage into ditches from a plane water table overlying a gravel substratum. *J. Geophys. Res.* **65**, 1267–1272 (1960).

24. Kirkham, D., Potential flow into circumferential openings in drain tubes. *J. Appl. Phys.* **21**, 655–660 (1950).
25. Ernst L. F., Grondwaterstromingen in de verzadigde zone en hun berekening bij aanwezigheid van horizontale evenwijdige open leidingen (Groundwater flow in the saturated zone and its calculation when horizontal parallel open conduits are present). Versl. Landbouwk. Onderz. No. 67.15, 189 pp. Wageningen, 1962.
26. Flodquist, H., Kulturtechnische Grundwasserforschungen. *Sveriges Geol. Undersökn.* **25** (1931).
27. MacKenzie, A. J., Chemical treatment of mineral deposits in drain tile. *J. Soil Water Conserv.* **17(3)**, 124–125 (1962).
28. Van Deemter, J. J., Bijdragen tot de kennis van enige natuurkundige grootheden van de grond, 11, Theoretische en numerieke behandeling van ontwatering en infiltratie-stromingsproblemen. *Verslag. Landbouwk. Onderzoek.* **56(7)**, 1950.
29. Engelund, F., Mathematical discussion of drainage problems. *Trans. Dan. Acad. Techn. Sci.* **3**, 64 (1951).
30. Kirkham, D., Steady-state theories for drainage. *Proc. Am. Soc. Civil Engrs.* **92(IR1)**, 19–39 (1966).
31. Vedernikov, V. V., Sur la théories du drainage. *Compt. Rend. (Doklady). Acadm. Sci. l'URRS, Moscow* **23**, 335–339 (1939).
32. List, E. J., The steady flow of precipitation to an infinite series of tile drains above an impervious layer. *J. Geophys. Res.* **69(16)**, 3371–3381 (1964).
33. van Schilfgaarde, J., Kirkham, D., and Frevert, R. K., Physical and mathematical theories of tile and ditch drainage and their usefulness in design. *Iowa Agr. Expt. Sta. Res. Bul.* 436, 667–706 (1956).
34. Kirkham, D., and Powers, W. L., An exact theory of seepage of steady rainfall into tile or ditch drained land. *Trans. Intern. Congress for Soil Sci., 9th, Bucharest*, (1964).
35. Kirkham, D., Seepage of steady rainfall through soil into drains. *Trans. Am. Geophys. Un.* **39(5)**, 892–908 (1958).
36. Toksöz, S., and Kirkham, D., Graphic solution and interpretation of a new drain-spacing formula. *J. Geophys. Res.* **66(2)**, 509–516 (1961).
37. Wesseling, J., Comparison of the steady state drain spacing formulas of Hooghoudt and Kirkham in connection with design practice. *J. Hydrol.* **2(1)**, 25–32 (1964).
38. Wesseling, J., The effect of using continually submerged drains on drain spacings. *J. Hydrol.* **2(1)**, 33–43 (1964).
39. DeWiest, R. J. M., History of the Dupuit-Forchheimer assumptions on groundwater hydraulics. *Trans. Am. Soc. Agr. Engrs.* **8(4)**, 508–516, 519 (1965).
40. Forchheimer, Ph., "Hydraulik." B. G. Teubner, Leipzig, 1914.
41. Polubarinova-Kochina, P. Ya., "Theory of Ground Water Movement" (J. M. R. DeWiest, translator). Princeton Univ. Press, Princeton, New Jersey, 1962.
42. Colding, L. A., Om lovene for vandets bavaeglse i jorden, Danske Vidensk. Selsk. Skr. 5. Raekke, Natur v. og math. Afd. 9B, 563–621. 1872 (Copenhagen).
43. Hooghoudt, S. B., Bijdragen tot de kennis van eenige natuurkundige grootheden van den grond, 6, Bepaling van de doorlatendheid in gronden van de tweede soort; theorie en toepassing van de kwantitatieve strooming van het water in ondiep gelegen grondlagen, vooral in verband met ontwaterings- en infiltratievraagstukken. *Verslag. Landbouwk. Onderzoek.* **42**, 461–676. Algemeene Landsdrukkerij, The Hague (1937).
44. Hooghoudt, S. B., Bijdragen tot de kennis van eenige natuurkundige grootheden van den grond, 7, Algemeene beschouwing van het probleem van de detail ontwatering en de infiltratie door middel van parallel loopende drains, greppels, slooten en kanalen. *Verslag. Landbouwk. Onderzoek* **46**, 515–707. Algemeene Landsdrukkerij, The Hague (1940).

45. Van Schilfgaarde, J., Design of tile drainage for falling water tables. *Proc. Am. Soc. Civil Engrs.* **89(IR2)**, 1–11 (1963).
46. Moody, W. T., Nonlinear differential equation of drain spacing. *Proc. Am. Soc. Civil Engrs.* **92(IR2)**, 1–9 (1966).
47. Guyon, G., Considérations sur l'hydraulique du drainage des nappes. Thesis. Toulouse, France, 1965.
48. Dagan, G., Spacing of drains by an approximate method. *Proc. Am. Soc. Civil Engrs.* **90(IR1)**, 41–66 (1964).
49. Childs, E. C., The nonsteady state of the water table in drained land. *J. Geophys. Res.* **65(2)**, 780–782 (1960).
50. Boussinesq, J., Sur le débit, en temps de secheresse, d'une source alimentée par une nappe d'eaux d'infiltration. *Compt. Rend. Scéances Acad. Sci.* **136**, 1511–1517. Gauthier-Villars, Paris (1903).
51. Dumm, L. D., New formula for determining depth and spacing of subsurface drains in irrigated lands. *Agr. Engr.* **35**, 726–730 (1954).
52. Krayenhoff van de Leur, D. A., A study of nonsteady groundwater flow with special reference to a reservoir-coefficient. *De Ingenieur* **70B**, 87–94 (1958).
53. Maasland, M., Water table fluctuations induced by intermittent recharge. *J. Geophys. Res.* **64**, 549–559 (1959).
54. Dumm, L. D., Transient-flow concept in subsurface drainage: Its validity and use. *Trans. Am. Soc. Agr. Engrs.* **7(2)**, 142–146, 151 (1964).
55. Werner, P. W., Some problems in non-artesian groundwater flow. *Trans. Am. Geophys. Un.* **38(4)**, 511–518 (1957).
56. Van Schilfgaarde, J., Closing discussion, Design of tile drainage for falling water tables. *Proc. Am. Soc. Civil Engrs.* **90(IR3)**, 71–73 (1964).
57. Brooks, R. H., Unsteady flow of groundwater into drain tile. *Proc. Am. Soc. Civil Engrs.* **87(IR2)**, 27–37 (1961).
58. Kirkham, D., Physical artifices and formulas for approximating water table fall in tile-drained land. *Proc. Soil Sci. Soc. Am.* **28(5)**, 585–590 (1964).
59. Dagan, G., Linearized solution of unsteady deep flow toward an array of horizontal drains. *J. Geophys. Res.* **69(16)**, 3361–3369 (1964).
60. Bouwer, H., and van Schilfgaarde, J., Simplified method of predicting fall of water table in drained land. *Trans. Am. Soc. Agr. Engrs.* **6(4)**, 288–291, 296 (1963).
61. De Zeeuw, J. W., and Hellinga, F., Neerslag en afvoer. *Landbouwk. Tijdschr.* **70**, 405–422 (1958).
62. Van Schilfgaarde, J., Transient design of drainage systems. *Proc. Am. Soc. Civil Engrs.* **91(IR3)**, 9–22 (1965).
63. Wiser, E. H., and van Schilfgaarde, J., Prediction of irrigation water requirements using a water balance. *Proc. Intern. Congress Agr. Engr., 6th*, 121–129 (1964).
64. Sewell, J. I., and van Schilfgaarde, J., Digital computer solutions of partially unsaturated steady-state drainage and subirrigation problems. *Trans. Am. Soc. Agr. Engr.* **6(4)**, 292–296 (1963).
65. Bouwer, H., Theoretical aspects of flow above the water table in tile drainage of shallow homogeneous soils. *Proc. Soil Sci. Soc. Am.* **23(4)**, 260–263 (1959).
66. Gardner, W. R., Some steady-state solutions of the unsaturated moisture flow equation with application to evaporation from a water table. *Soil Sci.* **85**, 228–232 (1958).
67. Brooks, R. H., and Corey, A. T., Hydraulic properties of porous media. Colo. State Univ. Hydrol. Paper No. 3, 1964.
68. Childs, E. C., A treatment of the capillary fringe in the theory of drainage. *J. Soil Sci.* **19(1)**, 83–100 (1959).

69. CHILDS, E. C., A treatment of the capillary fringe in the theory of drainage II. *J. Soil Sci.* **11(2)**, 293–304 (1960).
70. MYERS, L. E., and VAN BAVEL, C. H. M., Measurement and evaluation of water table elevations. *Congr. Intern. Comm. Irrigation Drainage 5th*, 17. 109–17. 119, 1962.
71. BOUWER, H., Unsaturated flow in groundwater hydraulics. *Proc. Am. Soc. Civil Engrs.* **90(HY5)**, 121–144 (1964).
72. GARDNER, W. R., Approximate solution of a nonsteady state drainage problem. *Proc. Soil Sci. Soc. Am.* **26(2)**, 129–132 (1962).
73. MASON, D. D., LUTZ, J. F., and PETERSEN, R. G., Hydraulic conductivity as related to certain soil properties in a number of Great Soil Groups—sampling errors involved. *Proc. Soil Sci. Soc. Am.* **21(5)**, 554–560 (1957).
74. KIRKHAM, D., Measurement of the hydraulic conductivity of the soil in place. *in* Symposium on Permeability of Soils, ASTM Spec. Tech. Pub. 163, pp. 80–97. 1955.
75. KIRKHAM, D., Saturated conductivity as a characterizer of soil for drainage design. *Proc. Am. Soc. Agr. Engr. Conf.*, "Drainage for Efficient Crop Production," pp. 24–31, 1965.
76. BOUWER, H., Field measurement of saturated hydraulic conductivity in initially unsaturated soil. *Intern. Assoc. Sci. Hydrol. Symp.*, Haifa, pp. 243–251 (1967).
77. BOUWER, H., Rapid field measurement of air entry value and hydraulic conductivity of soil as significant parameters in flow system analysis. *Water Resources Res.* **2(4)**, 729–738 (1966).
78. BOUWER, H., Developing design requirements for parallel drains. *Proc. Am. Soc. Agr. Engr. Conf.*, "Drainage for Efficient Crop Production." pp. 62–65. 1965.
79. REEVE, R. C., Characterizing soil properties for drainage design—Art or Science? *Proc. Am. Soc. Agr. Engr. Conf.*, "Drainage for Efficient Crop Production." pp. 21–23, 1965.
80. DUMM, L. D., and WINGER, R. J., Subsurface drainage system design for irrigated area using transient-flow concept. *Trans. Am. Soc. Agr. Engr.* **7(2)**, 147–151 (1964).
81. VAIGNEUR, H. O., and JOHNSON, H. P., Drainage design based on moisture balance. *Trans. Am. Soc. Agr. Engr.* **9(6)**, 764–767 (1966).
82. FRENCH, B. E., and O'CALLAGHAN, J. R., A field test of drain spacing equations for agricultural land. *J. Agr. Engr. Res.* **11(4)**, 282–295 (1966).

SEA WATER

R. A. HORNE*

Arthur D. Little, Inc.
Cambridge, Massachusetts

I. Introduction

A. The Oceans as a Major Feature of Earth

If something in outer space is watching our planet, the spectacle must be splendid, for the most conspicuous feature of Earth to the distant observer is probably the brilliant reflection of the sun's rays off the vast sheet of liquid water that covers slightly more than 70% of the planet's surface, or some 36,000,000 km^2. The mean depth of the Earth's hydrosphere is nearly

* Present address: Woods Hole Oceanographic Institution, Woods Hole, Massachusetts.

3,800 m. The quantity of liquid in this mantle is enormous, 1,400,000,000 km^3, representing more than 1,400,000,000,000,000,000 kg of sea water [*1*].

B. The Role of Sea Water in Biogenesis

Granted that the oceans are the most obvious fact of our world, there is a second, much more profound, highly special reason why sea water is important to us; why an understanding of it should be central to all our efforts to unravel the physical world about us. The chemistry of life is the chemistry of aqueous electrolytic solutions. To such an extent that we can safely declare that where there is no water there is no life, and wherever in the cosmos there is adequate liquid water there may be living forms. What is so special about water? Liquid water is easily the most special substance known to man, it is, as I have repeatedly emphasized elsewhere [*2–4*] anomalous in every single one of its physical–chemical properties. Little wonder then, that the most special of phenomena, life, should have appeared and evolved in that most special of media, sea water. If time has left any trace of the innermost secrets of biogenesis and evolution those traces are to be found in the seas. As our knowledge of the structure of water, biopolymer hydration, and hydrogen and hydrophobic bonding slowly grows a marvelous picture is emerging of how the first building blocks of the protobiotic substance were put together millions of years ago [*3*, Ch. 13]. Even when they left the oceans, the living things brought with them in their body fluids their original marine environment and in the periodic rhythm of their life functions the pulse of ancient tides. Sometimes the elemental composition of the body fluids gives us clues to the evolutionary history of the species. For example, the common bony fish such as cod and pollack have low Mg/Ca ratios in their blood serum, comparable to land and fresh-water animals rather than to the high ratios characteristic of more primitive marine species, indicating that they were once fresh-water forms that returned to the oceans in more recent geological time [*5*].

C. The Oceans and Our Total Environment

We have seen that sea water hides many of the keys to our past; it also contains the keys to our future . . . if the human race has a future. We are a part of the biosphere of this planet. The perturbation of the atmosphere, lithosphere, and hydrosphere by biopopulations is not negligible. On the contrary, the atmosphere, for example, is secondary, that is to say not that of the primeval Earth, and its large oxygen content is the result of biological activity—specifically photosynthesis. The oceans of Earth are neither an inexhaustible cupboard nor an infinite sewer. The growing human population is making ever-increasing demands upon sustenances and water resources while at the same time diminishing the usefulness of these resources by contamination and pollution. At last there now appears to be some awakening to the seriousness and enormity of these problems and there is a growing concern, both scientific and lay, with the rapid deterioration of our total environment. Consequently we can expect increased scientific interest in the major components of that ecology—hydrosphere, lithosphere, atmosphere, and biosphere—and their interactions. Here again the physical chemistry of sea water is crucial. Can, for example, the world's oceans absorb the carbon dioxide poured into the atmosphere by human activity and absorb it fast enough in order to avert the catastrophic consequences of a gradual warming up of the atmosphere?

D. The Marine Environment

In this chapter we are going to restrict our attention to, as it were, an element of representative sea water sampled from the oceans. Now in the oceans that element of sea water will be subjected to certain environment conditions, and in this section I would like to examine briefly the more significant of these variables—namely, temperature and pressure. The temperature of the surface waters ranges from about $-2°$ to $+30°C$ and can fluctuate widely in both space and time exhibiting diurnal and seasonal variations as well as more erratic changes arising from local oceanographic and meteorological conditions. Then with increasing depth the temperature commonly begins to fall off gradually. This region of the vertical thermal profile of the water column (Fig. 1) is known as the *main thermocline* and it often exhibits a seasonal fluctuation. The temperature of the deep water is quite constant, lying in the range of $-1°$ to $+4°C$. The deep ocean waters are an excellent thermostat.

The hydrostatic pressure increases with increasing depth at the rate of about 1 decibar per meter. Figure 2 indicates the percentage of the Earth's ocean areas of a given depth range along with the corresponding pressure. At the bottom of the most profound depths in the oceans, such as the Philippine Trench, the pressure can be as much as 1000 atm. The deep ocean environment, then, is characterized by low and very constant temperature and high, hydrostatic pressure. This, as cosmic environments go, is a most extraordinary one, and the effects of temperature and pressure must be taken into careful consideration by those who examine the physical–chemical processes and chemical equilibria occurring in sea water in the deep ocean.

Generally speaking, our specimen of sea water will be moving as the results of waves, tides, ocean currents, upwelling, or sinking. Such gross mass transfer is the subject matter of physical oceanography and lies outside the compass of this article.

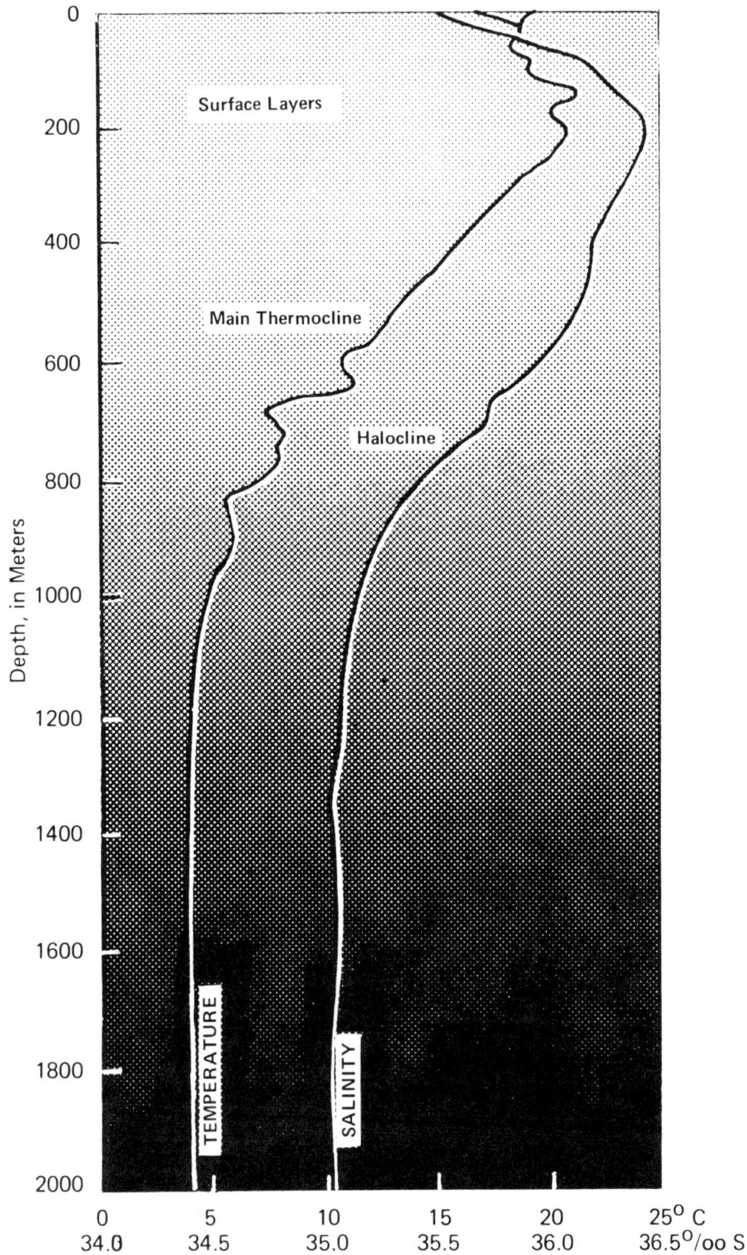

FIG. 1. Typical thermal and salinity profiles in the oceans.

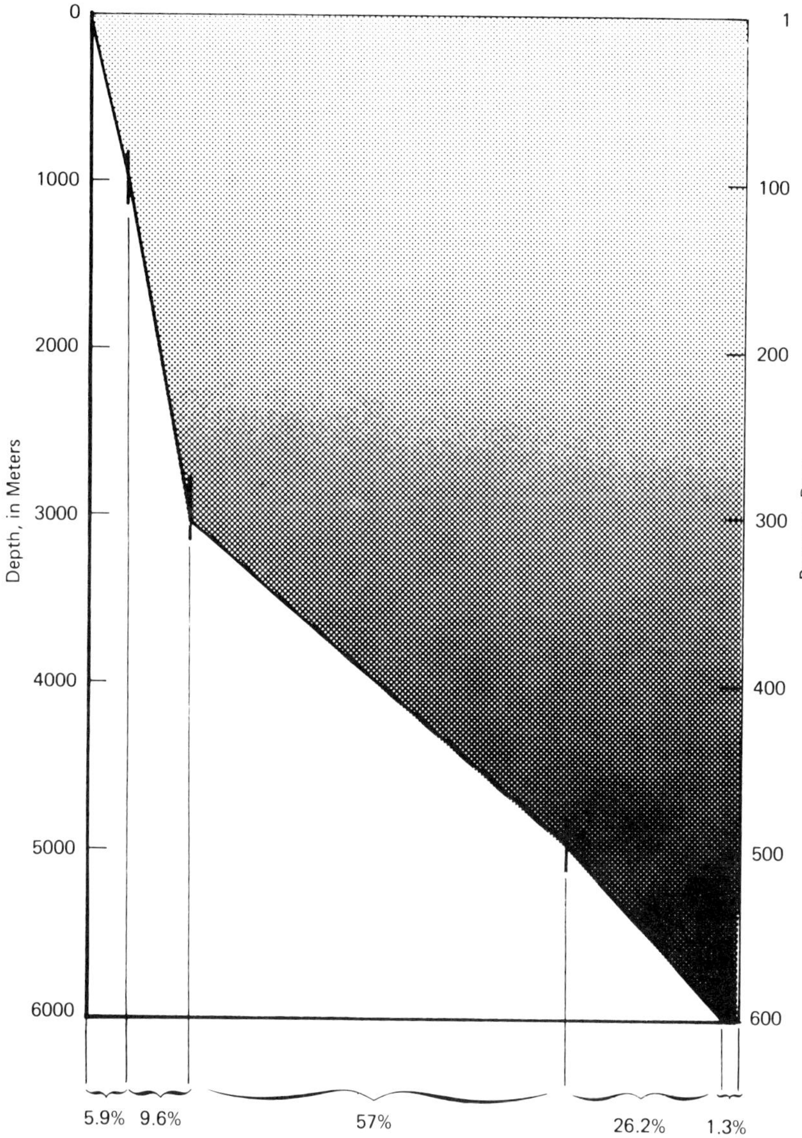

FIG. 2. Percentage of the ocean areas in a given depth–pressure range.

II. The Constituents of Sea Water

A. Classification of the Substances in Sea Water

Sea water is roughly an aqueous $0.5M$ NaCl solution, approximately $0.05M$ in $MgSO_4$, and containing a trace of just about everything else imaginable. And I do mean literally everything from gold to DNA, noble gases to radioisotopes, old shoes to submarines, viruses to whales. These countless substances can be lumped into categories on any of a number of bases. I feel that the classification shown in Fig. 3 is both an obvious and

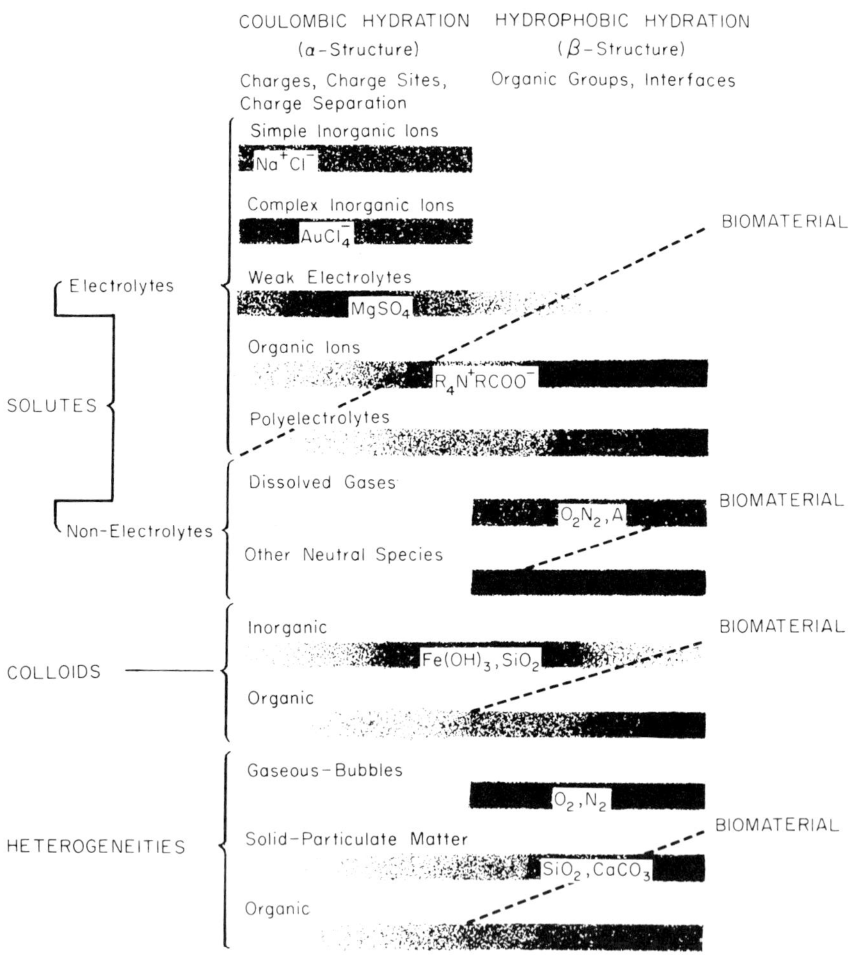

FIG. 3. Classification of substances in sea water [4].

useful one, both practically and theoretically. In the left-hand column in the figure the substances in sea water are divided into two categories: the solutes —those forming one phase with and dissolved in the sea water—and those present as heterogeneities, or a second phase. A given substance can fall into both categories; N_2 from the atmosphere, for example, can both dissolve in sea water or be entrapped in bubbles. The dissolved substances in turn are further divided into electrolytes, or those which upon dissolution dissociate to form charged species capable of carrying an electrical current, and non-electrolytes, or neutral species, which do not. The distinction between truly dissolved substances and heterogeneities is not a sharp one. Practically it is usually defined by some sort of physical filtration procedure. Bridging these two classes is a third ill-defined but extremely important class—the *colloids*. Our knowledge of the role that colloidal substances play in the chemistry of fresh water, sea water, and estuaries is in its infancy.

The second and third columns in Fig. 3 make reference to a fundamental difference in the nature of the water envelopes surrounding the substances and are, at the present time, primarily of theoretical significance. Alternatively, we might have classified the substances in sea water as inorganic and organic; but, as is so often the case in nature, such a dichotomy is not a very satisfactory one for purely inorganic substances, such as $CaCO_3$, which can be very important participants in complex biochemical processes.

B. The Constancy of Species Ratios

The atmosphere of the Earth is well mixed. As a consequence the relative amounts of its major constituents, N_2 and O_2, are relatively invariable. Similarly, although the total solute content varies widely from place to place and temporally, the *ratios* of the dissolved constituents in sea water remain *nearly* constant, indicating that the hydrosphere too, while a much more viscous medium, is nevertheless also relatively well mixed. (Table I).

TABLE I

Major Constituent Ratios for Selected Oceans and Seas

	Na^+Cl^a	Mg^{++}/Cl^a	K^+/Cl^a	Ca^{++}/Cl^a	$SO_4^=/Cl^a$
Atlantic Ocean	0.5544	0.0667	0.0195	0.0212	0.1393
Pacific Ocean	0.5553	0.0663	0.0209	0.0215	0.1396
Baltic Sea	0.5536	0.0669	—	0.0216	0.1414
Mediterranean Sea	0.5528	0.0679	0.0201	—	0.1396

[a] Cl is the chlorinity as defined in the text.

In the words of Marcet (1819), "All the specimens of sea-water contain the same ingredients all over the world, these bearing very nearly the same proportions to each other, so that they differ only as to the total amount of their saline contents." This near constancy of the relative amounts of the solutes in sea water has four consequences which can be put to good use: (1) it enables the meaningful compilation of *a* table of the elemental composition of sea water, (2) it enables us to define a useful parameter—the salinity—describing the total salt content of sea-water, (3) it enables a large segment of marine chemistry to be described in terms of the variation of that parameter only, and (4), and by no means the least important, the exceptions to the rule constitute and deliniate an important class of solutes—the nutrient substances, that is to say, the "inorganic" ions such as NO_3^-, NO_2^-, HCO_3^-, $H_2PO_4^-$, NH_4^+, etc., which are crucial in the cycles of living organisms in the sea.

C. The Elemental Composition of Sea Water

First let us turn to the concentration of all of the known chemical elements in sea water. A chemical science follows a logical sequence of historical development in five steps of increasing difficulty:

1. Identification and quantitative determination of the chemical elements present in the system

↓

2. Determination of the specific chemical form in which the elements are present—i.e., chemical equilibria

↓

3. Resolution of the kinetics of those equilibria

↓

4. Elucidation of the mechanism of the physical–chemical processes

↓

5. Extension of the foregoing to heterogeneous systems and boundary processes

and marine chemistry is by no means exempt from this scheme of things. Prior to World War II, chemical oceanography was largely a servant to the other marine disciplines and much of its activity was devoted to the detection and measurement of the chemical elements present in sea water. Although this work still continues, especially with the application of more sophisticated analytical instrumentation, the principal occupation of marine chemists at the present time is the identification of the oxidation states of the elements and of chemical equilibria in the oceans.

Table II lists the elemental composition of sea water as given by Goldberg [*6*] with his later values for the rare-earth elements [*7*] added. This table also summarizes the principal chemical species of the element in sea water and the elements' residence time in the oceans. The analytical chemistry of sea water has been reviewed in detail by Barnes [*8*], Culkin [*9*], and Riley [*10*].

For purposes of laboratory physical–chemical measurements the International Association of Physical Oceanography, Charlottenlund Slot, Denmark, operates a standard sea-water service that provides sealed ampules of sea water of chlorinity (see below) specified to five significant figures. Alternatively, artificial sea water can be prepared from any of a number of recipes given in the literature. One of the most recently published [*11*] of these is given in Table III.

D. Salinity Variations in the Oceans

Above we mentioned as one of the advantages resulting from the constancy of the ratios of sea-water constituents, the utility of characterizing sea water in terms of its total salt content or *salinity*, S. The salinity (expressed in grams of solid per kilogram of sea water, ‰) is related to the chlorinity, Cl, by the expression

$$S = 0.03 + 1.805 \text{ Cl} \tag{1}$$

Although this latter term is both a nuisance and an anachronism (it is based on the classical wet Volhard analysis for chloride and other halogens), oceanographers persist in using it and continue, consequently, to harvest confusion. We might notice immediately that S is still 0.03 when Cl is equal to zero, hence expression (1) is clearly not applicable to very dilute sea-water solutions. For all its inexactness, however, the salinity, as a measure of total salt content, does appear to be of sufficient utility to warrant its further use and Cox and his co-workers [*12–15*] have established a good correlation between the salinity of sea water, the density, and the electrical conductivity—a property easily measured to a very high degree of accuracy.

Turning now to the salinity variations in the oceans, they are for the most part exactly what one would expect, in particular the salinity of the surface water depends strongly on the balance between evaporation and condensation

TABLE II

THE ELEMENTAL COMPOSITION OF SEA WATER

Element	Abundance, mg/l	Principal species	Residence time years
H	108,000	H_2O	
He	0.000005	He(g)	
Li	0.17	Li^+	2.0×10^7
Be	0.0000006		1.5×10^2
B	4.6	$B(OH)_3$; $B(OH)_2O-$	
C	28	HCO_3-; H_2CO_3; CO_3^{2-}; organic compounds	
N	0.5	NO_3-; NO_2-; NH_4+; $N_2(g)$; organic compounds	
O	857,000	H_2O: $O_2(g)$; SO_4^{2-} and other anions	
F	1.3	F−	
Ne	0.0001	Ne(g)	
Na	10,500	Na^+	2.6×10^8
Mg	1350	Mg^{2+}; $MgSO_4$	4.5×10^7
Al	0.01		1.0×10^2
Si	3	$Si(OH)_4$; $Si(OH)_3O-$	8.0×10^3
P	0.07	HPO_4^{2-}; H_2PO_4-; PO_4^{3-}; H_3PO_4	
S	885	SO_4^{2-}	
Cl	19,000	Cl−	
Ar	0.6	Ar(g)	
K	380	K^+	1.1×10^7
Ca	400	Ca^{2+}; $CaSO_4$	8.0×10^6
Sc	0.00004		5.6×10^3
Ti	0.001		1.6×10^2
V	0.002	$VO_2(OH)_3^{2-}$	1.0×10^4
Cr	0.00005		3.5×10^2
Mn	0.002	Mn^{2+}; $MnSO_4$	1.4×10^3
Fe	0.01	$Fe(OH)_3(s)$	1.4×10^2
Co	0.0005	Co^{2+}; $CoSO_4$	1.8×10^4
Ni	0.002	Ni^{2+}; $NiSO_4$	1.8×10^4
Cu	0.003	Cu^2; $CuSO_4$	5.0×10^4
Zn	0.01	Zn^{2+}; $ZnSO_4$	1.8×10^5
Ga	0.00003		1.4×10^3
Ge	0.00007	$Ge(OH)_4$; $Ge(OH)_3O-$	7.0×10^3
As	0.003	$HAsO_4^{2-}$; H_2AsO_4-; H_3AsO_4; H_3AsO_3	
Se	0.004	SeO_4^{2-}	
Br	65	Br −	
Kr	0.0003	Kr(g)	
Rb	0.12	Rb+	2.7×10^5
Sr	8	Sr^{2+}; $SrSO_4$	1.9×10^7
Y	0.0003		7.5×10^3
Zr			
Nb	0.00001		3.0×10^2

TABLE II *(Continued)*

Element	Abundance, mg/l	Principal species	Residence time years
Mo	0.01	MoO_4^{2-}	5.0×10^5
Tc			
Ru			
Rh			
Pd			
Ag	0.00004	$AgCl_2^-$; $AgCl_3^{2-}$	2.1×10^6
Cd	0.00011	Cd^{2+}; $CdSO_4$	5.0×10^5
In	<0.02		
Sn	0.0005		1.0×10^5
Sb	0.0005		3.5×10^5
Te			
I	0.06	IO_3^-; I^-	
Xe	0.0001	Xe(g)	
Cs	0.0005	Cs^+	4.0×10^4
Ba	0.03	Ba^{2+}; $BaSO_4$	8.4×10^4
La	1.2×10^{-5}		4.4×10^2
Ce	5.2×10^{-6}		8.0×10^5
Pr	2.6×10^{-6}		3.2×10^2
Nd	9.2×10^{-6}		2.7×10^2
Pm			
Sm	1.7×10^{-6}		1.8×10^2
Eu	4.6×10^{-7}		3.0×10^2
Gd	2.4×10^{-7}		2.6×10^2
Tb			
Dy	2.9×10^{-6}		4.6×10^2
Ho	8.8×10^{-7}		5.3×10^2
Er	2.4×10^{-6}		6.9×10^2
Tm	5.2×10^{-7}		1.8×10^3
Yb	2.0×10^{-6}		5.3×10^2
Lu	4.8×10^{-7}		4.5×10^2
Hf			
Ta			
W	0.0001	WO_4^{2-}	1.0×10^3
Re			
Os			
Ir			
Pt			
Au	0.000004	$AuCl_4^-$	5.6×10^5
Hg	0.00003	$HgCl_3^-$; $HgCl_4^{2-}$	4.2×10^4
Tl	<0.00001	Tl^+	
Pb	0.00003	Pb^{2+}; $PbSO_4$	2.0×10^3
Bi	0.00002		4.5×10^5
Po			
At			

TABLE II *(Continued)*

Element	Abundance, mg/l	Principal species	Residence time, years
Rn	0.6×10^{-15}	Rn(g)	
Fr			
Ra	1.0×10^{-10}	Ra^{2+}; $RaSO_4$	
Ac			
Th	0.00005		3.5×10^2
Pa	2.0×10^{-9}		
U	0.003	$UO_2(CO_3)_3^{4-}$	5.0×10^5

TABLE III

PREPARATION OF ARTIFICIAL SEA WATER

23.926 gm NaCl 4.008 gm Na_2SO_4 0.677 gm KCl 0.196 gm $NaHCO_3$ 0.098 gm KBr 0.026 gm H_3BO_3 0.003 gm NaF 53.27 ml of a 1.0*M* $MgCl_2$ solution 10.33 ml of a 1.0*M* $CaCl_2$ solution 10.96 ml of a 0.5*M* $SrCl_2$ solution	Dilute with distilled water to make a total weight of 1000 gm, equilibrate, aerate, and, if necessary adjust the *p*H to between 7.9 and 8.3

processes [*3*, Ch. 5]. Evaporation during the day may increase the salinity from its darkness value of say 35.46‰ to, say, 35.50‰, while seasonal variations in the North Atlantic range from a high of 36.70 in March to a low of 36.00 in November. The salinity of the surface waters decreases sharply during rainfall, but, due to mixing processes, it resumes its initial value within 2–3 hours after precipitation ceases [*16*]. Perhaps the most spectacular salinity variation is the dependence of the global variations of surface salinity on latitude. In the tropical regions (20°N and 20°S) high temperatures and strong trade winds give such high evaporation rates that salinity maxima in the 35.7‰ range are observed. Towards the far North and South, the salinity then falls off below 34.5‰ as rainfall exceeds evaporation, and at the equator a minimum of about 34.5‰ appears as a result of increased precipitation and reduced wind velocities. The salinity displays a vertical profile reminiscent of the thermal structure (Fig. 1) with highly variable surface values, highly constant deep values, and an intermediate region of gradually changing salinity sometimes called the *halocline*. Temperature and salinity profiles are highly characteristic of specific regions and seasons, so much so that plots of temperature versus salinity can be used to identify and follow the movement of water masses in the oceans.

Ice formation will tend to increase the salinity of the sea water in a region, whereas melting ice will tend to dilute the sea water. The influx of fresh waters from rivers will also lower the salinity of sea water. The lighter fresh water can continue to flow over the heavier sea water for great distances without mixing resulting in a salinity profile in which, in contrast to Fig. 1, salinity increases with increasing depth in the halocline. Somewhat the opposite of this situation occurs outside the Straits of Gibraltar where the heavier more saline water from the Mediterranean Sea (37–38‰ S) flows down the western slope of the Trafalgar Rise underneath the less saline (36‰ S) Atlantic Ocean waters.

Highly anomalous salinities can result from particular local conditions: for example, hot (45°C), very saline (>43‰ S), sodium- and calcium-enriched waters discovered at the bottom of the Red Sea presumably are connate waters formed by leaching processes in a sedimentary rock sill [*17–20*], while on the other extreme, low-salinity water-flows off the coast of Sicily may have submarine fresh water springs as their source [*21*].

E. Marine Equilibria

One of the most important advances in marine chemistry [*22*] in recent years has been the postulation of chemical equilibrium models by Garrels and Thompson [*22*] and Sillen [*23–25*]. The concentrations of the major species present in 19‰ Cl sea water at 25°C and 1 atm calculated by Garrels and Thompson [*22*] are summarized in Table IV; they also give the activities

TABLE IV

Principal Species Present in 19 ‰ Cl Sea Water at 25°C and 1 atm[a]

Element					
Sodium	0.4752 m	99% Na^+	1.2% $NaSO_4^-$	0.01% $NaHCO_3$	—
Magnesium	0.0540	87% Mg^{++}	11% $MgSO_4$	1% $MgHCO_3^+$	0.3% $MgCO_3$
Calcium	0.0104	91% Ca^{++}	8% $CaSO_4$	1% $CaHCO_3^+$	0.2% $CaCO_3$
Potassium	0.0100	99% K^+	1% $NaSO_4^-$	—	—
Anionic Species					
Sulfate	0.0284	54% $SO_4^=$	3% $CaSO_4$	21.5% $MgSO_4$	21% $NaSO_4^-$
Bicarbonate	0.0024	69% HCO_3^-	4% $CaHCO_3^+$	19% $MgHCO_3^+$	8% $NaHCO_3$
Carbonate	0.0003	9% $CO_3^=$	7% $CaCO_3$	67% $MgCO_3$	17% $NaCO_3^-$

[a] According to the Chemical Model of Garrels and Thompson [*22*].

of the free ions as 3.56×10^{-1}, 1.69×10^{-2}, 6.3×10^{-3}, 2.64×10^{-3}, 1.79×10^{-3}, 9.75×10^{-4}, and 4.7×10^{-6} for Na^{+}, Mg^{++}, K^{+}, Ca^{++}, $SO_4^{=}$, HCO_3^{-}, and $CO_3^{=}$, respectively.

In order to determine the specific species in which a given element is largely present in sea water, one must first determine the valence state in the oxidative or reducing environment represented by the waters in question. The redox potential of sea water is sometimes expressed in terms of *p*E, analogous to *p*H. Sillen [*23*] takes the value of *p*E to be 12.5 ± 0.2. And he further points out that this parameter is, fortunately, relatively insensitive to the usual small changes in *p*H and oxygen concentration, it being necessary to reduce the latter by 40% of its saturation value in order to lower *p*E by 0.1 unit. There are, however, extensive areas in the oceans and seas where anoxic or reducing conditions prevail, and these zones have a chemistry quite different from that of "normal" sea water [*26*]. Next, inasmuch as many equilibria involve hydrogen ion the *p*H must be known in addition to the concentrations of the element and the other species with which it is likely to combine. Finally, the concentrations of the most likely species are estimated by using the appropriate equilibrium constants. Such constants at the proper temperature, ionic strength, and more especially at high pressure are rarely available; however, usually a satisfactory semiquantitative estimate can be made on the basis of the 25°C, 1-atm values.

Of the very great number of chemical equilibria in the seas, only two are mentioned here as examples: the magnesium sulfate and the carbonate system.

The overall equilibrium

$$Mg^{++} + SO_4^{=} \rightleftharpoons MgSO_4, \qquad K_{MgSO_4} = a_{MgSO_4}/a_{Mg^{++}}\, a_{SO_4^{=}} \tag{2}$$

where K is the *equilibrium constant* and the a's represent activities, is responsible for the anomalous attenuation of acoustic energy in sea water. Because of the importance of this absorption to undersea warfare, the temperature and pressure dependence of this equilibrium, unlike most marine reactions, have been carefully studied [*26–28*]. As a consequence of electrostriction, that side of an equilibrium of greatest charge will be the state of smaller volume; therefore, the application of hydrostatic pressure forces reaction (2) to the left, that is to say, the dissociation of the weak electrolyte $MgSO_4$ increases with increasing pressure.

Apart from solubility studies, the pressure dependences of only a very few equilibria significant in marine chemistry have been measured. These exceptions include the dissociation of several weak acids [*29–31*], and of iron (III) chloride complex ion [*32*].

One of the most complex and certainly one of the most important equilibrium systems in the oceans is the carbonate system which is fundamental not only to the hydrosphere but to the atmosphere, lithosphere, and biosphere as well (Fig. 4). This system is believed to maintain the *p*H of the oceans so

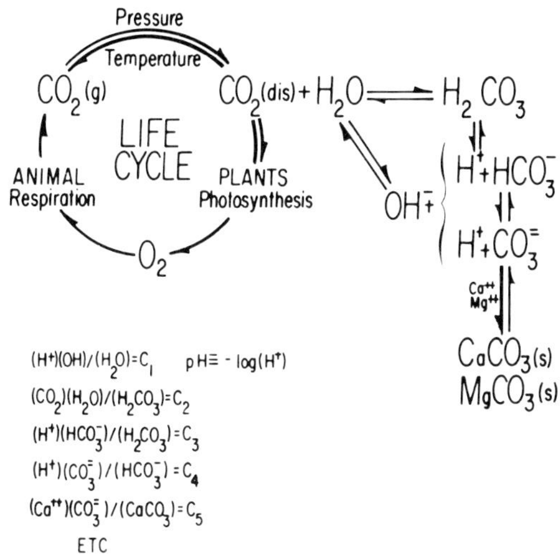

FIG. 4. Simplified schematic diagram of the carbonate system of equilibria [*3*].

close to neutrality, with perhaps some assistance from equilibria involving silicate minerals [*23*, *33*, *34*], and has been described elsewhere in greater detail by Skirrow [*35*] and myself [*3*, ch. 7]. Analysis of the equilibria shows that at the lower end of the marine *p*H range the principal dissolved form is HCO_3^- with a little CO_2 and at the upper end, mostly HCO_3^- with a little $CO_3^=$ (Fig. 5). For all the attention that has been paid to it, many unsolved

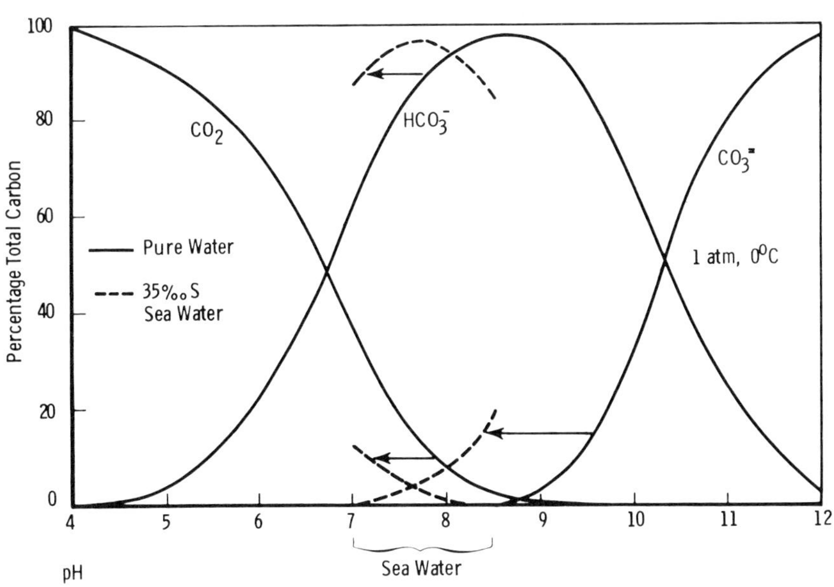

FIG. 5. The form of carbonate in pure water and sea water [*3*].

mysteries surround the carbonate system. The supersaturation in some regions and undersaturation in others of the carbon dioxide content of sea water, even if one takes into account the perturbation of the CO_2 concentration by photosynthesis, make little sense. A possible explanation is that the dissolution of atmospheric CO_2 into sea water is slow. This brings us to a fundamental difficulty of the proposed chemical models of sea water—the oceans are almost certainly *not* at equilibrium. Nevertheless, with respect to many systems they are at least close to equilibrium so that such models remain very useful. When such models work, we are pleased; and when they fail, we are not surprised. The dissolution of particulate $CaCO_3$ in sea water is also a strange business. Here not only may rate phenomena be involved again but the presence of organic matter may further impede dissolution [*36*]. We can expect the kinetics of marine processes to occupy the labors of an increasing number of oceanographers and environmental scientists.

F. Suspended Material

In our classification of the substances in sea water (Fig. 3) we distinguished between dissolved solutes and those present as a second phase or heterogeneity. This second phase can be in the form of gas bubbles or in the form of particulate solids (the second phase can also be an organism). The amount of suspended particulate matter in sea water is highly variable but it is always very much less than the amount of dissolved material; an Atlantic sample contained between 30–40 gm of dissolved salts but only 0.003 gm of solid material [*37*], and in clear water the amount can easily be an order of magnitude less. In addition to its influence on the optical properties suspended particulate matter (and especially bubbles) can cause strong scattering and absorption of acoustic energy thus degrading sonar performance.

The suspended material in turn can be further classified as inorganic and organic. The former is largely of a mineral nature. The suspended organic material is largely detritus and although its quantity is commonly larger than that of the inorganic material it is usually very much less than the total amount of dissolved organic material, for example 9 compared to 89% in the euphotic zone. The nature and concentration of the detritus is highly variable, especially in coastal areas.

As for the inorganic matter, silicon is an important constituent. The marine chemistry of this element is both very important and very complex. It occurs in both the dissolved, particulate, and colloidal state and it plays a central role in both geological and biological processes in the oceans [*38*; *3*, Ch. 8]. Atmospheric phenomena also should not be overlooked for great quantities of siliceous material finds its way into the ocean waters and the sediments beneath via airborne dust from the continental land masses [*39*].

G. The Organic Content of Sea Water

Earlier I said that sea water contains a trace of just about everything imaginable. While its total amount is smaller than the quantity of dissolved inorganic salts in sea water, the variety of the organic material is endless and, inasmuch as it represents the food chain of all the living creatures in the sea, its importance can hardly be overestimated. As we have just seen, some of the organic material is in particulate form, including intact living organisms, but more is dissolved. The biomaterial of marine organisms can be divided into three categories: (1) the living biosubstance proper composed of complex organic compounds, (2) the body fluids, and (3) the supporting or containing skeletal structures [*3*, Ch. 9]. The chemical elements found in the biosubstance fall in turn into two classes: (1) the biopolymeric structural elements, such as hydrogen, carbon, oxygen, nitrogen, phosphorous, sulfur, etc., present in appreciable amounts, and (2) the catalytic elements, such as copper, zinc, iron, vanadium, bromine, etc. found in trace amounts. The composition of the various body fluids tends to be very similar both with respect to electrolyte concentration and *p*H, to that of sea water. The skeletal material is composed largely of the carbonates, silicates, or phosphates of calcium and magnesium and, I repeat participates both in biological and geological cycles in the sea. When organisms die or decay, these substances or their decomposition products may be released into the sea water. Thus, in addition to the waste products of the life processes, one finds in sea water a complete spectrum of organic material ranging on the one hand from intact living organisms to, on the other extreme, the final simplest products of the decay process. One finds carbohydrates, lipids, proteins, enzymes, catalysts such as chlorophyll, and a host of amino acids—in short, all the reactants, all the products, and all the intermediates of the great biochemical processes which occur in the ocean: photosynthesis, animal metabolism, decomposition, and mineral fixation.

H. Nutrients (and Products)

There are a number of materials—O_2, CO_2, NO_3^-, NO_2^-, $PO_4^{\equiv}$, $SiO_4^=$, NH_4^+, which, while they are simple inorganic species, play a vital role in the life processes. The concentration of these substances at a given place and time is a direct function of the biological activity occurring in those waters and, as a result, they form a most significant exception to our earlier rule that the ratios of species in sea water remain constant. As the biopopulation grows, the concentration of these nutrients in the water will be diminished only to be replenished upon the death or dissolution of the organisms. This is illustrated with the silicon cycle in Fig. 6. The biopopulation of micro-

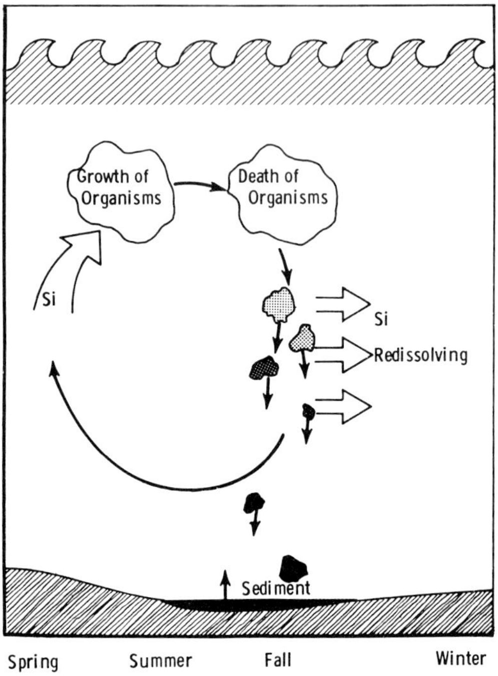

FIG. 6. The silicon cycle in the oceans [3].

organisms is the food basis of larger organisms, including fish of commercial potential, hence knowledge of the waxing or waning and movement of nutrient substances is of increasing practical concern.

These same nutrients are also of considerable theoretical interest to the oceanographer for they form a sort of biological clock which results in the chemical characterization of sea water depending on its age. The concentrations of the nutrients correlate well, not only with biological activity, but because of their biochemical relationships, with one another. The nutrient ratios tend to be different and characteristic of the world's several great oceans; and the vertical profiles for the distribution of phosphate, nitrate, or silicate in the Atlantic Ocean, for example, is quite different from that of the Pacific or Indian Oceans. As waters age the traces of biological activity slowly accumulate, and on the basis of such evidence we discover that the Atlantic waters are younger than the Pacific, and we can trace the northward movement of slowly aging waters in the Pacific Ocean.

I. GASES AND RADIOISOTOPES IN THE SEA

Table V summarizes the results of measurements of the solubilities of various gases in sea water while Table VI lists some *in situ* gas concentrations

TABLE V

SOLUBILITIES OF SOME GASES IN SEA WATER AT 1 ATM

Gas	Chlorinity	Temperature	Concentration	Reference
CO	15 ‰ Cl	2°C	0.02880[a] ml/ml	Douglas [40]
CO	20	2	0.02709[a]	
CO	15	10	0.02428[a]	
CO	20	10	0.02296[a]	
O_2	14 ‰ Cl	2°C	0.00819[b] ml/ml	Green and Carritt [41]
O_2	20	2	0.00760[b]	
O_2	14	10	0.00678[b]	
O_2	20	10	0.00633[b]	
N_2	15 ‰ Cl	2°C	0.01845[a] ml/ml	Douglas [43]
N_2	20	2	0.01720[a]	
N_2	15	10	0.01554[a]	
N_2	20	10	0.01463[a]	
Ar	15 ‰ Cl	2°C	0.04212[a] ml/ml	Douglas [43]
Ar	20	2	0.03968[a]	
Ar	15	10	0.03504[a]	
Ar	20	10	0.03312[a]	
CO_2	15‰ Cl	2°C	0.0623 moles/liter	Riley and Skirrow [49]
CO_2	20	2	0.0593	
CO_2	15	10	0.0472	
CO_2	20	10	0.0452	

[a] 760 mm of gas (STPD).
[b] Water-saturated atmosphere 0.2094 mole fraction in O_2.

in sea water. As a point of reference, the usage is to assume that every specimen of sea water has been at some time in its history at the sea's surface and in equilibrium with the atmospheric gases, and that, subsequently, its concentration has been altered by physical, chemical, and biological processes occurring in the sea. For example, O_2 and CO_2 are strongly perturbed by photosynthetic activity whereas Ar and to a lesser extent N_2 are more or less "inert" in the sea. Further details on gas solubilities and concentrations in sea water can be found in reviews by Richards [*53*], Skirrow [*35*], and myself [*3*].

The radioisotopes present in sea water come from three sources: (1) natural, long half-life nuclides of the uranium and thorium series decaying since the birth of this planet and their shorter half-life decay products, (2) natural, shorter half-life nuclides formed continually by processes such as cosmic radiation in the Earth's atmosphere, and (3) artificial nuclides from weapons testing and other human pollution [*54*, *55*].

Representative elements are ^{238}U, ^{230}Th, ^{226}Ra, and ^{231}Pa; ^{14}C, ^{3}H, ^{32}Si, and ^{10}Be; and ^{90}Sr, ^{137}Cs, ^{147}Pm, ^{157}Sm, ^{60}Co, ^{54}Mn, and ^{55}Fe, respectively

TABLE VI

CONCENTRATIONS OF SOME GASES IN SEA WATER

Gas	Location	Concentration	Reference
Methane, CH_4	Gulf of Mex., Surf.	0.676×10^{-7} ml/ml	Swinnerton and Linnenbom [42]
CH_4	Gulf of Mex., 500 m	0.246×10^{-7}	
CH_4	N. Atlantic, Surf.	0.474×10^{-7}	
CH_4	N. Atlantic, 500 m	0.488×10^{-7}	
He	Pacific, 46 m	0.44×10^{-7} ml/ml	Bieri, Koide, and Goldberg [44]
He	Pacific, 3000 m	0.50×10^{-7}	
Ar	Pacific, 46 m	0.31×10^{-7} ml/ml	
Ar	Pacific, 3000 m	0.29×10^{-7}	
N_2	Atlantic	10 to 108% of Sat.	Rakestraw and Emmel [45]
N_2 and Ar	Pacific	91 to 102% of Sat.	Hamm and Thompson [46]
O_2	S. Atlantic, Surf.	100 to 107% of Sat.	Truesdale, *et al.* [47]
O_2	Bay of Biscay	0.005 to 0.006 ml/ml	Cooper [48]
CO_2	Pacific	250 to 330 ppm	Waterman [50]
CO_2	Indian Ocean	270 to 330 ppm	
CO_2	Atlantic	290 to 340 ppm	
CO_2	Atlantic, 1 m	292 to 301 ppm	Takahasi [51]
CO_2	Atlantic, 10 m	299 to 301 ppm	
CO_2	S. Pacific, Surf.	300 ppm	Keeling, *et al.* [52]
CO_2	S. Pacific, 400 m	600 ppm	

in each category. The radioisotope ^{40}K, which decays by β emission and K-electron capture accounts for more than 90% of the natural radioactivity of sea water. The next most important contribution is ^{87}Rb, but its increment is less than 1%.

Like the biological aging mentioned above, radioisotopic aging provides an extremely valuable research tool for identifying and examining the history and movement of water masses in the oceans and the exchanges between the hydrosphere and the lithosphere and atmosphere [56]; ^{14}C has proven to be particularly useful in such studies.

J. THE ORIGINS OF THE OCEANS

This section concludes with a few brief words about the origins of the oceans. The water of the oceans of Earth did not come from the condensation of the hot primeval atmosphere of the planet, for, in the long period of geological time involved, the Earth's gravitational field is insufficient to prevent the loss of gaseous substances into space. The oceans of the Earth then must be of a secondary nature, that is to say the water and other volatiles found in the sea water must have been originally chemically bound in the lithosphere and subsequently released by plutonic processes after the planet had cooled [*57*]. And, in fact, calculations based on the output of existing thermal springs indicates that over a period of 3×10^9 years such sources can easily account for the total amount of water in the oceans and in the atmosphere, biosphere, and ancient sedimentary rocks as well [*57*].

The origins of the inorganic solutes in the sea is perhaps an even more fascinating story. If we compare the concentrations of elements in sea-water with what we might expect from the weathering and leaching of the Earth's crust, we are disappointed to find seeming chaos. However, on more careful examination, taking sodium, which has so little chemistry in the sea, as a sort of standard, we find the solutes fall into three classes: (1) those that, like Na, have more or less the expected concentration in sea water, (2) those in very much smaller concentrations, and (3) those with concentrations much larger than expected. The elements apparently depleted in sea water are those such as Si and Al, which form highly insoluble compounds, those such as Fe, Mn, Ni, and V, which are known to be scavanged out of sea water by precipitation, absorption, and concretion processes, and those such as Ca, Mg, and C (in the form of $CO_3^=$) removed and fixed by biological processes. The elements present in much greater amounts than they should be—S, Cl, Br, and B—are all elements that form highly volatile compounds. This led Rubey [*57*] to the theory of excess volatiles: while the cations in sea water come largely from the weathering and leaching of surface rocks, the anionic substances come by a different route (Fig. 7) and that route is not rivers, but rather volcanic emanations [*57*; see also *58–61*].

III. Physical–Chemical Properties of Sea Water

A. *PVT* Properties

Scandalously enough, many of the fundamental physical–chemical properties of sea water have never been accurately measured, especially in the low-temperature high-pressure range. For the physical chemist sea water is an unattractive soup. It is a sufficiently dilute solution so that with respect to solvent properties it is fairly ideal, departing from ideality as given by Raoult's law by only about 0.4%. On the other hand, with the respect to solute, it is far too concentrated a solution to be tractable by the elaborate theories that have been developed for treating ideal solutions. For example, the equivalent electrical conduction is about 26% less than the conductance of a corresponding infinitely dilute solution [*3*, Ch. 2].

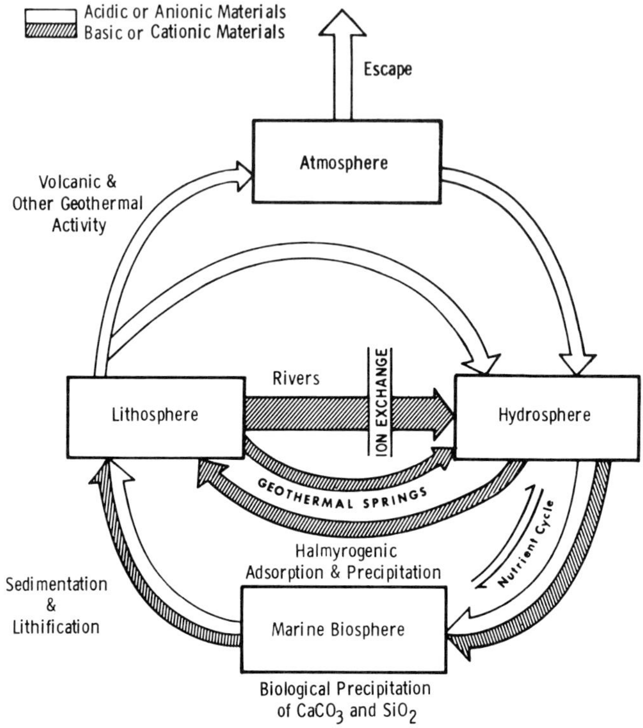

FIG. 7. The origins of anionic and cationic solutes in the oceans of earth [*3*].

Reliable thermodynamic data on sea water are largely lacking, and the density is estimated as a function of salinity, pressure, and temperature on the basis of empirical equations or read from hydrographic tables. The accuracy of these tables has been discussed by Eckart [*62*], Crease [*63*], and Newton and Kennedy [*64*]. These deficiencies have been recognized, and progress is being made in filling in the more conspicuous information gaps. Wilson and Bradley [*65*], for example, have made what appear to be very careful measurements of the compressibility and coefficient of thermal expansion of sea water as a function of salinity, temperature, and pressure. These measurements illustrate an instance where the scientific community should be most grateful for a military need. The specific volume data in Table VII are abstracted from their work. The temperature of maximum density of sea water at 1 atm also depends on the salinity decreasing from its pure-water value of 3.947°C to 1.860, −0.310, and −2.473°C for 10, 20, and 30‰ S sea water.

Several attempts have been made to formulate an equation of state for sea-water but the results, for the most part, have been purely empirical.

TABLE VII

SPECIFIC VOLUME OF SEA WATER (cm^3/gm)

		0°C	10°C	20°C
Pure water	1 bar	1.0000	1.0003	1.0017
	500 bars	0.9767	0.9781	0.9804
	1000 bars	0.9567	0.9590	0.9618
10‰ Cl	1 bar	0.9921	0.9926	0.9942
	500 bars	0.9695	0.9711	0.9735
	1000 bars	0.9501	0.9526	0.9554
20‰ Cl	1 bar	0.9842	0.9850	0.9868
	500 bars	0.9623	0.9642	0.9666
	1000 bars	0.9435	0.9461	0.9491
30‰ Cl	1 bar	0.9764	0.9775	0.9794
	500 bars	0.9552	0.9572	0.9598
	1000 bars	0.9370	0.9397	0.9428
40‰ Cl	1 bar	0.9687	0.9770	0.9720
	500 bars	0.9481	0.9504	0.9530
	1000 bars	0.9305	0.9334	0.9365

The recent equation proposed by Li [*66*] may prove to be of interest as it is based on the Tait–Gibson relation which in turn may be related to the structure of liquid water [*67*].

The velocity of sound V in sea water is given (in meters per second) by the empirical equation of Wilson [*68*]:

$$V = 1449.22 + \Delta V_T + \Delta V_P + \Delta V_S + \Delta V_{STP} \tag{3}$$

The values of the coefficients are given in Table VIII.

TABLE VIII

COEFFICIENTS IN THE SOUND VELOCITY EQUATION

$T = -3°C$	0°C	5°C	10°C	20°C
$\Delta V_T = 14.37$	0.00	+21.79	+41.05	+72.81
$P = 1$ bar	200 bars	400 bars	600 bars	800 bars
$\Delta V_P = +0.16$	+32.64	+65.98	+100.30	+135.33
$S = 33‰$ S	34‰ S	35‰ S	36‰ S	37‰ S
$\Delta V_S = -3.09$	−1.47	−0.00	+1.31	+2.47
$P = 1$ bar	200 bars	400 bars	600 bars	800 bars
$\Delta V_{STP} = 0.00$[a]	+0.31	+0.73	+1.30	+2.05

[a] At 35‰ S.

B. Colligative and Transport Properties

The presence of a solute alters the thermodynamic activity of the solvent thereby changing certain of its properties. Table IX shows the dependency of some of these colligative properties of sea water on salinity. Recently

TABLE IX

Colligative Properties of Sea Water
Osmotic Pressure P, Freezing Point ΔT_{FP}, Boiling-Point Elevation ΔT_{BP}, and Vapor-Pressure Lowering VP_V as a Function of Salinity[a]

S‰	P_0, atm	P_{20}, atm	ΔT_{FP}, °C	ΔT_{BP}, °C	P_V, mm Hg
4	2.59	2.78	−0.214	0.06	1.69
8	5.16	5.54	−0.427	0.12	3.39
12	7.73	8.30	−0.640	0.19	5.08
16	10.34	11.10	−0.856	0.25	6.82
20	12.97	13.92	−1.074	0.31	8.55
24	15.63	16.78	−1.294	0.38	10.30
28	18.31	19.65	−1.516	0.44	12.13
32	21.02	22.56	−1.740	0.51	13.97
36	23.76	25.50	−1.967	0.57	15.79
40	26.53	28.48	−2.196	0.64	17.73

[a] From Dietrich [*5*].

several papers have appeared in the literature reporting more detailed measurements on the colligative properties of sea water, synthetic sea water, and solutions of sea-water constituents: these include new tables of the boiling-point elevation by Fabuss and Korosi [*69*] and Stoughton and Lietzke [*70*] and vapor pressure and osmotic data by Robinson [*71*], Aron and Kientzler [*72*], and Stoughton and Lietzke [*70*].

Table X gives some selected values of the specific heat of sea water at 1 atm from the work of Cox and Smith [*73*].

In the sea, except near interfaces, ion and molecule movement is largely by macroscopic mixing processes rather than by microscopic transport processes such as diffusion. Nevertheless the ion and molecule transport processes in sea water are of interest in their own right [*3*, Ch. 3] and for analysis of the physical–chemistry of processes occurring at the boundaries of the marine environment—at the air–sea interface and in marine sediments. Particular attention has focused on the electrical conductivity of sea water. Next to the temperature, it is perhaps the most frequently measured property. It is a property which can be measured at sea with relatively great ease and

TABLE X

SPECIFIC HEAT OF SEA WATER[a]

	−2°C	0°C	5°C	15°C	25°C
0‰ S	—	4.217	4.202	4.186	4.179
10‰ S	—	4.142	4.135	4.128	4.126
20‰ S	—	4.074	4.072	4.071	4.072
30‰ S	—	4.013	4.014	4.016	4.020
40‰ S	3.957	3.957	3.959	3.965	3.972

[a] In abs. J/gm.

accuracy and it appears to correlate directly with the total salt content or salinity and the density. In view of these advantages the proposal has been repeatedly made that the electrical conductivity, rather than the chlorinity, be specified as the primary characterization of the composition of sea water [*74*]. The electrical conductivity of sea water at 1 atm is shown in Fig. 8. Even though sea water is a relatively concentrated solution, ***Walden's rule*** appears still to provide a good approximation of its conductive properties at 1 atm, thus for 35‰ S sea water in going from 0° to 20°C the conductance changes by 61% and the viscosity by 49% but their product only by 5% [*75*]. The effect of hydrostatic pressure on the electrical conductivity of sea water

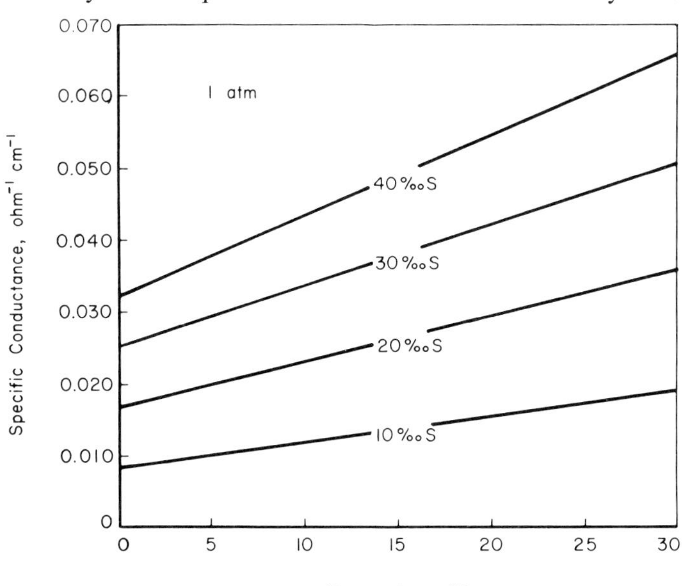

FIG. 8. The electrical conductivity of sea water at 1 atm [*3*].

has been examined experimentally by Horne and Frysinger [*76*] and Bradshaw and Schleicher [*77*]. At 25°C the conductivity of 35‰ Cl sea water is about 6% greater at 1000 bars than at 1 atm and at 0°C nearly 12% greater. These increases are greater than expected from the volume and viscosity changes and the increased dissociation of the weak electrolyte $MgSO_4$ and have been attributed to the dehydration of ions with a consequent reduction in their effective radii in solution [*75*] (see also below).

Another way of stating Walden's rule is to say that the viscosity varies as the reciprocal of the electrical conductance. As the temperature increases the conductance increases but the viscosity decreases (Table XI, values selected

TABLE XI

RELATIVE VISCOSITY OF SEA WATER AT 1 ATM[a]

	10‰ S	20‰ S	30‰ S	40‰ S
0°C	1.017	1.032	1.056	1.054
10°C	0.745	0.758	0.772	0.785
20°C	0.574	0.586	0.599	0.611
30°C	0.460	0.470	0.481	0.491

[a] Relative to pure water at 0°C, $\eta = 1.787$ cP.

from the compilation of Dorsey [*78*]). Notice also that the viscosity increases as the concentration of solute increases. The viscosity of sea water decreases with increasing pressure (Fig. 9) but the minimum in the viscosity versus pressure curve does not exactly correspond to the maximum in the conductance versus pressure curve indicating that other processes (such as ion dehydration) are going on.

The self-diffusion coefficient for pure water is 1.46, 1.68, 1.97, and 2.57×10^{-5} cm^2/sec at 1 atm and 5°, 10°, 15°, and 25°C, respectively [*80*] and presumably is not very different for sea water. The ionic diffusion coefficient for 0.5*M* NaCl is 1.474×10^{-5} cm^2 sec^{-1} at 25°C while that of $MgSO_4$ (0.005*M*) is much less, 0.710×10^{-5} cm^2 sec^{-1} [*81*], Richardson *et al.* [*82*] tried to measure the diffusion of NaCl in sea water under pressure. The experimental difficulties were considerable, and the investigators concluded that pressure has little effect on diffusion—a rather surprising finding.

IV. The Structure of Sea Water

A. PURE WATER

The objectives of science extend beyond the description of natural phenomena, for the goal of science is understanding, and we do not under-

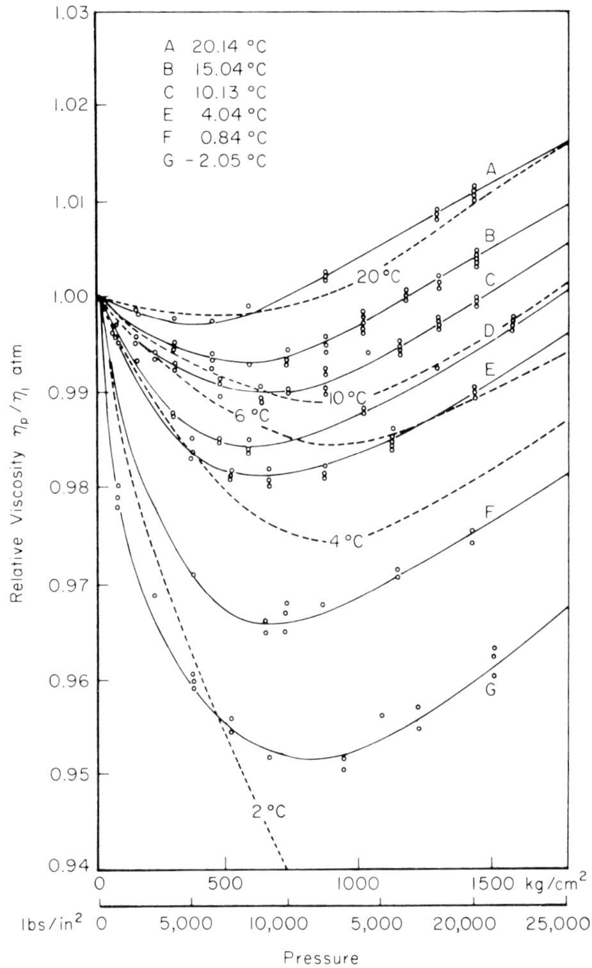

FIG. 9. Relative viscosity of 19.375 ‰ Cl sea water under pressure (dotted areas are pure water) [*79*].

stand until we have addressed the question why? as well as how? Oceanography in the past has been descriptive. The time is at hand to take a further step forward and attempt to explain the phenomena described in terms of molecular processes and forces (step 4 of our schedule above). We must, so to speak, abandon the muralist's brush for the anatomist's dissection needle. Accordingly, the final section of this article is devoted to reviewing the molecular anatomy of sea water. This is a topic very close to my own immediate area of interest and one that, accordingly, I have dealt with repeatedly and in greater detail elsewhere [*2–4*, *83*].

There are many theories of the structure of liquid water and the subject has been a highly controversial one for more than thirty years, particularly in recent years with a quickening of interest in water on the part of ecologists, oceanographers, physiologists, and molecular biologists [*84–87*]. Of these many viewpoints my own preference continues to be in favor of the "flickering cluster" theory by Frank and Wen [*88*]. Liquid water, according to this model is composed of a "mixture" of constantly forming, constantly dissolving (half-life 10^{-10} to 10^{-11} sec) aggregates of hydrogen-bonded water molecules and *more or less* "free" or monomeric water (Fig. 10). Because of the co-

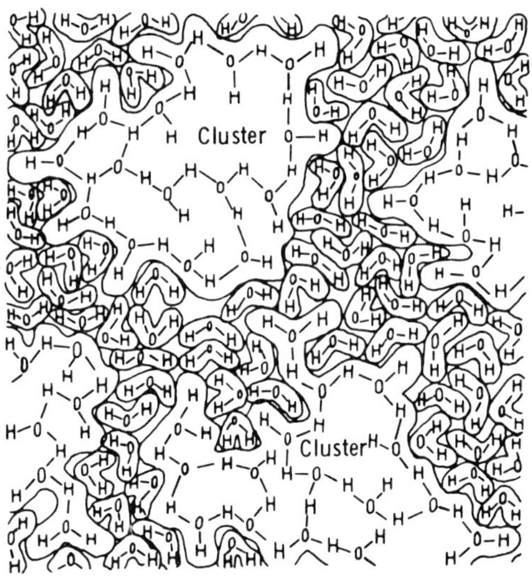

FIG. 10. Diagrammatic model of the structure of pure liquid water [*4*].

operative nature of the hydrogen bonding process, in contrast to the smaller polymers proposed by pre-1933 theories of water structure, the Frank–Wen clusters are quite large: Nemethy and Scheraga [*89*] estimate that at 20°C the average cluster contains about 40 water molecules. It should be cautioned that although referred to as "icelike" the structure of the clusters probably is *not* that specifically of hexagonal Ice I and that the nonclustered waters are *not* "free" in the sense of molecules in the vapor phase.

B. SEA WATER

The addition of a solute to water profoundly alters the structure of that liquid, especially in the immediate neighborhood of the solute, to form a hydration atmosphere. At least two essentially different types of hydration atmosphere appear to be formed—*hydrophobic hydration* in the case of nonpolar and *Coulombic hydration* for polar or ionic solutes [*4*]. Figure 3 classifies the substances in sea water with respect to their hydration type. The structure of the Coulombic hydration envelope is quite complex. Figure 11 diagrams

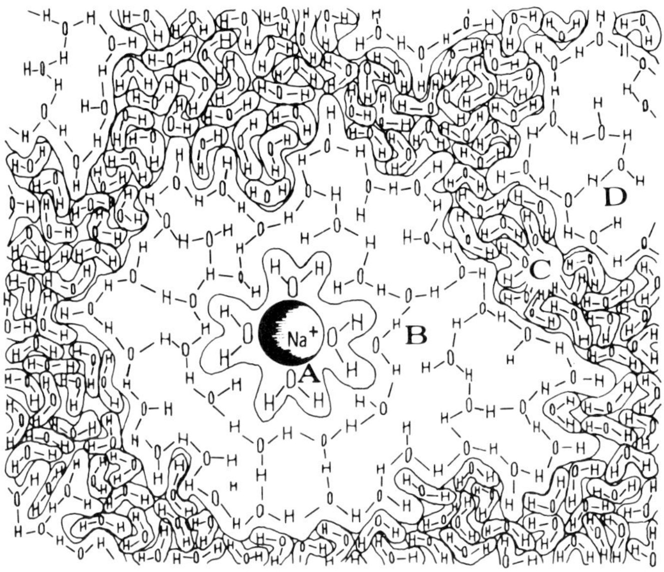

FIG. 11. Water structure near a sodium ion [*4*].

the structure surrounding a sodium ion in sea water. Immediately adjacent to the ion is its "primary hydration" sheath of 4 strongly bound, tightly electrostricted water molecules (11A), this is in turn surrounded by what is in essence a Frank–Wen cluster (11B). Beyond the region of enhanced water structure is a zone (11C) of broken water structure where the Coulombic field is strong enough to disrupt the normal structure of liquid water yet insufficiently potent to reorient the water molecules into some new pattern. At some considerable distance from the ion (11D) the structure of water resumes its "normal" fraction of hydrogen bonding. At 20°C the number of water molecules in the total hydration atmosphere of the Na^+, that is to say 11A plus 11B, is about 40—the same as in an average Frank–Wen cluster at the same temperature [*90*]. Little is known about the hydration of anions but in general they are believed to be less heavily hydrated than cations.

C. The Effects of Temperature and Pressure

With increasing temperature the Frank–Wen cluster fragment and dissolve, but the effect is a comparatively mild one, so that liquid water retains an appreciable fraction of its hydrogen-bonded character even at 100°C. Similarly, the operancy of Walden's rule for sea water indicates that temperature has little effect on the Coulombic hydration envelopes of Na^+ and Cl^- in sea water, or at least upon that fraction of an ion's hydration envelope that moves when the ions move.

The H-bonded clusters are less dense than the "free" water, hence the application of pressure destroys them and the viscosity decreases (Fig. 9). By about 1000 bars all of the structured regions have been destroyed so the viscosity goes through a minimum and then begins to increase with increasing pressure as a "normal" unassociated liquid should [*79*]. In the same way the application of pressure destroys the clusterlike region surrounding ions [Fig. 11(B)] and, if great enough so that the density of the bulk water exceeds that of the electrostricted region [Fig. 11(A)], even may strip away all or part of the primary hydration sheath. The changes in the structure of sea water when submitted to 5,000 gm/cm^2 are summarized in Fig. 12.

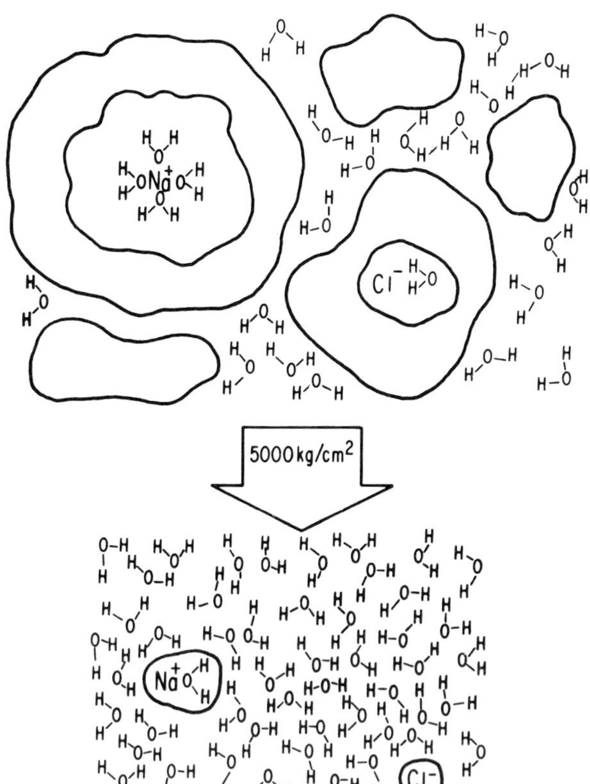

FIG. 12. The effect of hydrostatic pressure on the structure of sea water [*3*].

REFERENCES

1. SVERDRUP, H. U., JOHNSON, M. W., and FLEMING, R. H., "The Oceans." Prentice-Hall, Englewood Cliffs, New Jersey, 1942.
2. HORNE, R. A., The structure of water and aqueous solutions. *Surv. Progr. Chem.* **4**, 1–43 (1968).
3. HORNE, R. A., "Marine Chemistry." Wiley (Interscience), New York, 1969.
4. HORNE, R. A., Effect of structure and physical characteristics of water on water chemistry, *in* "Chemical and Microbiological Analysis in Water and Water Pollution" (L. L. Ciaccio, ed.). Dekker, New York, 1970 (in press).
5. DIETRICH, G., "General Oceanography." Wiley (Interscience), New York, 1963.
6. GOLDBERG, E. D., The oceans as a chemical system, *in* "The Sea" (M. N. Hill, ed.), Vol. 2, Chapter 1. Wiley (Interscience), New York, 1963.
7. GOLDBERG, E. D., Minor elements in sea water, *in* "Chemical Oceanography" (J. P. Riley and G. Skirrow, eds.), Vol. 1, Chapter 5. Academic Press, New York, 1965.
8. BARNES, H., "Apparatus and Methods of Oceanography, 1, Chemical." Allen & Unwin, London, 1959.
9. CULKIN, F., The major constituents of sea water, *in* "Chemical Oceanography" (J. P. Riley and G. Skirrow, eds.), Vol. 1, Chapter 4. Academic Press, New York, 1965.
10. RILEY, J. P., Analytical chemistry of sea water, *in* "Chemical Oceanography" (J. P. Riley and G. Skirrow, eds.), Vol. 2, Chapter 21, Academic Press, New York 1965.
11. KESTER, D. R., DUEDALL, I. W., CONNORS, D. N., and PYTKOWICZ, R. M., Preparation of artificial seawater. *Limnol. Oceano.* **12**, 176–179 (1967).
12. COX, R. A., CULKIN, F., GREENHALGH, R., and RILEY, J. P., Chlorinity, conductivity, and density of sea-water. *Nature* **193**, 518–520 (1962).
13. COX, R. A., Definition of the "salinity" of sea-water. *Nature* **197**, 785–786 (1963).
14. COX, R. A., The salinity problem. *Progr. Oceano.* **1**, 243–261 (1963).
15. COX, R. A., CULKIN, F., and RILEY, J. P., The electrical conductivity-chlorinity relationship in natural seawater. *Deep-Sea Res.* **14**, 203–220 (1967).
16. DEFANT, A., "Physical Oceanography." Pergamon Press, Oxford, 1961.
17. MILLER, A. R., High salinity in sea water. *Nature* **203**, 590–591 (1964).
18. SWALLOW, J. C., and CREASE, J., Hot salty water at the bottom of the Red Sea. *Nature* **205**, 165–166 (1965).
19. NEUMANN, A. C., and CHAVE, K. E., Connate origin proposed for hot salty bottom water from a Red Sea basin. *Nature* **206**, 1346–1347 (1965).
20. CRAIG, H., Isotopic composition and origin of the Red Sea and Salton Sea geothermal brines. *Science* **154**, 1544–1548 (1966).
21. CASSINIS, R., The detection of fresh-water submarine flows by means of *in situ* salinity measurements of sea-water. *Ric. Sci.* **37**, 267–280 (1967).
22. GARRELS, R. M., and THOMPSON, M. E., A chemical model for sea water. *Am. J. Sci.* **260**, 57–66 (1962).
23. SILLEN, L. G., The physical chemistry of sea water, *in* "Oceanography" (M. Sears, ed.), No. 67, pp. 549–584, Am. Assoc. Adv. Sci., Publ., Washington, D.C., 1961.
24. SILLEN, L. G., The ocean as a chemical system. *Science* **156**, 1189–1197 (1967).
25. SILLEN, L. G., Gibbs phase rule and marine sediments, *in* "Equilibrium Concepts in Natural Water Systems" (W. Stumm, ed.), Ser. No. 67, Chapter 4, Am. Chem. Soc. *Advan. Chem.*, Washington, D.C., 1967 (see also Ch. 3).
26. RICHARDS, F. A., Anoxic basins and fjords, *in* "Chemical Oceanography" (J. P. Riley and G. Skirrow, eds.), Vol. 1, Chapter 13. Academic Press, New York, 1965.
27. FISHER, F. H., The effect of pressure on the equilibrium of magnesium sulfate. *J. Phys. Chem.* **66**, 1607–1611 (1962).

28. Fisher, F. H., Multistate dissociation and the effect of pressure on the equilibrium on the magnesium sulfate. *J. Phys. Chem.* **69**, 695–696 (1965).
29. Disteche, A., Electromchemical measurements at high pressures. *J. Electrochem. Soc.* **109**, 1084–1092 (1962).
30. Disteche, A., and Disteche, S., The effects of pressure on *p*H and dissociation constants. *J. Electrochem. Soc.* **112,** 350-354 (1965).
31. Disteche, A., and Disteche, S., The effect of pressure on the dissociation of carbonic acid. *J. Electrochem. Soc.* **114**, 330–340 (1967).
32. Horne, R. A., Myers, B. R., and Frysinger, G. R., The effect of pressure on the dissociation of iron (III) monochloride complex ion in aqueous solution, *Inorg. Chem.* **3**, 452–454 (1964).
33. Garrels, R. M., Silica: role in buffering natural waters. *Science* **148**, 69 (1965).
34. Mackenzie, F. T., and Garrels, R. M., Silicates: reactivity with sea water. *Science* **150**, 57–58 (1965).
35. Skirrow, G., The dissolved gases—carbon dioxide, *in* "Chemical Oceanography" (J. P. Riley and G. Skirrow, eds.), Vol. 1, Chapter 7. Academic Press, New York, 1965.
36. Chave, K. E., Carbonates: association with organic matter in surface seawater. *Science* **148**, 1723–1724 (1965).
37. Groot, J. J., and Ewing, M., Suspended clay in a water sample from the deep ocean. *Science* **142**, 579–580 (1963).
38. Armstrong, F. A. J., Silicon, *in* "Chemical Oceanography" (J. P. Riley and G. Skirrow, eds.), Vol. 1, Chapter 10. Academic Press, New York, 1965.
39. Arrhenius, G., Sedimentary record of long-period phenomena, *in* "Advances in Earth Science" (P. M. Hurley, ed.), pp. 155–174. MIT Press, Cambridge, Massachusetts, 1966.
40. Douglas, E., Carbon monoxide solubilities in sea water. *J. Phys. Chem.* **71**, 1931–1933 (1967).
41. Green, E. J., and Carritt, D. E., New tables for oxygen saturation in seawater. *J. Marine Res.* (*Sears Found. Marine Res.*) **25**, 140–147 (1967).
42. Swinnerton, J. W., and Linnenbom, V. J., Gaseous hydrocarbons in sea water. *Science* **156**, 1119–1120 (1967).
43. Douglas, E., Solubilities of argon and nitrogen in sea water. *J. Phys. Chem.* **69**, 2608–2610 (1965).
44. Bieri, R., Koide, M., and Goldberg, E. D., Noble gases in sea water. *Science* **146**, 1035–1037 (1964).
45. Rakestraw, N. W., and Emmel, V. M., The relation of dissolved oxygen to nitrogen in some Atlantic waters. *J. Marine Res.* (*Sears Found. Marine Res.*) **1**, 207–216 (1938).
46. Hamm, R. E., and Thompson, T. G., Dissolved nitrogen in the sea water of the northwest Pacific. *J. Marine Res.* (*Sears Found. Marine Res.*) **4**, 11–27 (1941).
47. Truesdale, G. A., Downing, A. L., and Lowden, G. F., The solubility of oxygen in pure water and sea water. *J. Appl. Chem.* **5**, 53–62 (1955).
48. Cooper, L. H. N., Vertical and horizontal movements in the ocean, *in* "Oceanography" (M. Sears, ed.), Pub. No. 67, pp. 599–621. Am. Assoc. Adv. Sci., Washington, D.C., 1961.
49. Riley, J. P., and Skirrow, G. (eds.), "Chemical Oceanography," Vol. 1, Appendix V. Academic Press, New York, 1965.
50. Waterman, L. S., Carbon dioxide in surface waters. *Nature* **205**, 1099–1100 (1965).
51. Takahashi T., Carbon dioxide in the atmosphere and in the Atlantic Ocean. *J. Geophys. Res.* **66**, 477–494 (1961).
52. Keeling, C. D., Rakestraw, N. W., and Waterman, L. S., Carbon dioxide in surface waters of the Pacific Ocean. *J. Geophys. Res.* **70**, 6087–6097 (1965).

53. RICHARDS, F. A., Dissolved gases other than carbon dioxide, *in* "Chemical Oceanography" (J. P. Riley, and G. Skirrow, eds.), Vol. 1, Chapter 6. Academic Press, New York, 1965.
54. BURTON, J. D., Radioactive nuclides in sea water, marine sediments, and marine organisms, *in* "Chemical Oceanography" (J. P. Riley and G. Skirrow, eds.), Vol. 2, Chapter 22. Academic Press, New York, 1965.
55. MIYAKE, Y., Artificial radioactivity in the sea, *in* "The Sea" (M. N. Hill, ed.), Vol. 2, Chapter 3. Wiley (Interscience), New York, 1963.
56. BROECKER, W., Radioisotopes and large-scale oceanic mixing, *in* "The Sea" (M. N. Hill, ed.), Vol. 2, Chapter 4. Wiley (Interscience), New York, 1963.
57. RUBEY, W. W., Geologic history of sea water. *Bull. Geol. Soc. Am.* **62**, 1111–1147 (1951).
58. CONWAY, E. J., Mean geochemical data in relation to oceanic evolution. *Proc. Roy. Irish Acad.* **48B**, 119–159 (1942).
59. HOLLAND, H. D., The history of ocean water and its effect on the chemistry of the atmosphere. *Proc. Natl. Acad. Sci. USA* **53**, 1173–1183 (1965).
60. REVELLE, R., On the history of the oceans. *J. Marine Res.* (*Sears Found. Marine Res.*) **14**, 446–461 (1955).
61. NICHOLLS, G. D., The geochemical history of the oceans, *in* "Chemical Oceanography" (J. P. Riley and G. Skirrow, eds.), Vol. 2, Chapter 20, Academic Press, New York, 1965.
62. ECKART, C., The equation of state of water and sea water. *Am. J. Sci.* **256**, 225–240 (1958).
63. CREASE, J., The specific volume of sea water under pressure as determined by recent measurements of sound velocity. *Deep-Sea Res.* **9**, 209–213 (1962).
64. NEWTON, M. S., and KENNEDY, G. C., An experimental study of the PVTS relations of seawater. *J. Marine Res.* (*Sears Found. Marine Res.*) **23**, 88–103 (1965).
65. WILSON, W., and BRADLEY, D., Specific volume, thermal expansion, and isothermal compressibility of sea-water. U.S. Naval Ord. Lab. Rept. No. NOL TR-66-103 (AD-635-120) (June 2, 1966).
66. LI, Y. H., Equation of state of water and sea water. *J. Geophys. Res.* **72**, 2665–2678 (1967).
67. GINELL, R., Derivation of the Tait equation and its relation to the structure of liquids. *J. Chem. Phys.* **34**, 1249–1252 (1961).
68. WILSON, W. D., Speed of sound in sea water as a function of temperature, pressure, and salinity. *J. Acoust. Soc. Am.* **32**, 641–644 (1960).
69. FABUSS, B. M., and KOROSI, A., Boiling point elevations of sea water and its concentrates. *J. Chem. Eng. Data* **11**, 606–609 (1966).
70. STOUGHTON, R. W., and LIETZKE, M. H., Thermodynamic properties of sea salt solutions. *J. Chem. Eng. Data* **12**, 101–104 (1967).
71. ROBINSON, R. A., The vapor pressure and osmotic equivalence of sea water. *J. Marine Biol. Assoc. U.K.* **33**, 449–455 (1954).
72. ARONS, A. B., and KIENTZLER, C. G., Vapor pressure of sea-salt solutions. *Trans. Am. Geophys. Union* **35**, 722–728 (1954).
73. COX, R. A., and SMITH, N. D., The specific heat of sea water. *Proc. Roy. Soc.* (*London*) **A252**, 51–62 (1959).
74. Conference on Physical and Chemical Properties of Sea-Water. Natl. Acad. Sci., U.S.A., Natl. Res. Council Publ. No. 600, Washington, D.C., 1959.
75. HORNE, R. A., and COURANT, R. A., Application of Walden's rule to the electrical conduction of sea water. *J. Geophys. Res.* **69**, 1971–1977 (1964).
76. HORNE, R. A., and FRYSINGER, G. R., The effect of pressure on the electrical conductivity of sea water. *J. Geophys. Res.* **68**, 1967–1973 (1963).

77. BRADSHAW, A., and SCHLEICHER, K. E., The effect of pressure on the electrical conductance of sea water. *Deep-Sea Res.* **12**, 151–162 (1965).
78. DORSEY, N. E., "Properties of Ordinary Water-Substance." Reinhold, New York, 1940.
79. HORNE, R. A., and JOHNSON, D. S., The viscosity of compressed seawater. *J. Geophys. Res.* **71**, 5275–5277 (1966).
80. WANG, J. H., Self-diffusion and structure of liquid water. *J. Am. Chem. Soc.*, **73**, 510–513 (1951).
81. DEVELL, L., Measurements of the self-diffusion of water in pure water, H_2O-D_2O mixtures, and solutions of electrolytes. *Acta Chem. Scand.* **16**, 2177–2188 (1962).
82. RICHARDSON, J. L., BERGSTAINSSON, P., GETZ, R. J., PETERS, D. L., and SPRAGUE, R. W., "Sea Water Mass Diffusion Coefficient Studies." Philco Aeronutronic Div. Pub., No. U-3021, W.O. 2053 (Dec. 1964), Office of Naval Res. Contract No. Nonr-4061 (00).
83. HORNE, R. A., The physical chemistry and structure of sea water. *Water Resources Res.* **1**, 263–276 (1965).
84. KAVANAU, J. L., "Water and Solute-Water Interactions." Holden-Day, San Francisco, California, 1964.
85. SAMOILOV, O. Y., "Structure of Aqueous Electrolyte Solutions and the Hydration of Ions." Consultants Bureau, New York, 1965.
86. WICKE, E., Structure formation and molecular mobility in water and in aqueous solutions. *Angew. Chem.* (Intern. Ed.) **5**, 106–122 (1966).
87. DROST-HANSEN, W., The structure of water and water-solute interactions, *in* "Equilibrium Concepts in Natural Water Systems" (W. Stumm, ed.), Chapter 5, pp. 70–120. Adv. Chem. Ser. No. 67, Amer. Chem. Soc., Washington, D.C., 1967.
88. FRANK, H. S., and WEN, W. Y., Structural aspects of ion-solvent interaction in aqueous solutions. *Discussions Faraday Soc.* **24**, 133–140 (1957).
89. NEMETHY, G., and Scheraga, H. A., A model for the thermodynamic properties of liquid water. *J. Chem. Phys.* **36**, 3382–3400 (1962).
90. HORNE, R. A., and BIRKETT, J. D., The total hydration atmospheres of the alkali-metal cations in aqueous solution. *Electrochem. Acta* **12**, 1153–1160 (1967).

TWO-LIQUID FLOWS IN POROUS MEDIA

JACOB BEAR*

Technion–Israel Institute of Technology
Haifa, Israel

* Parts of this manuscript were written during the academic year 1966/67 while the author was a Visiting Professor at the Massachusetts Institute of Technology, Cambridge, Massachusetts.

I. Introduction

The purpose of this chapter is to present some basic concepts, techniques, and methods of investigation related to problems involving the flow of two fluids, and especially two liquids, through a porous medium domain. Such problems are often encountered in ground water hydrology and (oil) reservoir engineering. Obviously, the discussion presented here is based on the theory of the flow of a homogeneous fluid in a porous medium. This theory is not repeated here, as it is assumed that the reader is familiar with it and with its applications. Occasionally, certain aspects of this theory are also discussed here, mainly for comparison purposes or in order to emphasize the special features involved in two-fluid flows.

Limited in extent, this is by no means a complete or comprehensive treatise of the subject. The objective of this chapter is to present only the basic elements of this subject and the methods by means of which solutions may be derived for problems of practical interest. For additional information, the reader is referred to the references cited here, and to many other articles published in professional publications. Of special interest are Refs. [*1–4*].

A. Types of Two-Fluid Flow

Several types of flow are possible when two or more fluids in motion occupy a porous medium domain.

Miscible displacement is a type of two-fluid flow in which the two fluids are completely soluble in each other. Capillary forces between the two fluids do not come into effect. The interfacial tension between the two fluids is zero, and the two fluids dissolve in each other. *A distinct fluid-fluid interface does not exist.* Often, especially in ground water hydrology, this case is also referred to as *hydrodynamic dispersion.* The main mechanisms causing mixing in this case are the microscopic velocity variations within the pore space and molecular diffusion. The treatment of this topic is beyond the scope of this chapter.

Immiscible displacement, or the simultaneous flows of two or more immiscible fluids (e.g., oil, water, and gas) in a porous medium domain. In this case, the interfacial tension between the two fluids is nonzero, and a distinct fluid-fluid interface separates the two fluids within each pore. A capillary pressure difference exists across the interface at each point on it. This type of flow is discussed in Sec. III.

B. The Abrupt Interface Approximation

Actually, since the fluids are either miscible or immiscible, an abrupt interface (in the macroscopic sense) between them, i.e., a continuous surface completely separating the two fluids, cannot exist. In miscible displacement, even when the two fluids (e.g., fresh water and salt water) are initially separated by an abrupt interface (a situation which is, in fact, possible only under laboratory conditions), a transition zone due to hydrodynamic dispersion is immediately created, across which the composition of the fluid varies from that of one fluid to that of the other fluid. The same is true for immiscible fluids. Moreover, in the latter case, due to capillary phenomena, a a complete displacement is impossible; the concepts of irreducible wetting fluid saturation and residual nonwetting fluid saturation (Secs. II, C and E) are related to this question.

Yet, in many cases of interest, the transition zone in the case of miscible fluids is relatively narrow, or the displacement of immiscible fluids is almost complete, such that for all practical purposes a fictitious abrupt interface may be assumed to separate the two fluids. On each side of this imaginary interface is a single phase only (or almost so). Although this is clearly an approximate approach, whenever justified it appreciably simplifies the treatment of the two-fluid problem on hand. This type of flow is treated in Secs. IV, V, and VI.

C. Occurrence

The various types of two-fluid flows in porous media mentioned before occur in many engineering problems.

In practically all oil reservoirs displacement mechanisms, especially in the process of production, the simultaneous flow of oil, water and gas is encountered. Of special interest is the *water drive* mechanism in which water partly or completely displaces oil and gas by intruding into the oil producing area along its boundaries. Simultaneous flow occurs also in *secondary recovery operations* in oil reservoirs where gas, air, or water are injected into a reservoir in order to increase the productivity of the reservoir after it has reached a state of substantially complete depletion of the initial content of energy required for oil expulsion. Of special interest in this respect is the case of *water flooding* where water is injected into the oil reservoir.

In oil reservoir engineering the miscible displacement type of flow, in the form of a secondary recovery process called *solvent drive*, is encountered also. In this process, the injected fluid is miscible with the displaced fluid. A mixed zone is created between the two fluids; behind the zone lies a region occupied only by the displacing fluid. For comparison, in the immiscible displacement case a zone with residual saturation of the displaced fluid lay behind the zone of simultaneous flow.

In *ground water hydrology*, which deals essentially with water, only displacements of the first type are encountered, namely, miscible displacements. The most known case is that of sea-water intrusion into coastal aquifers. Sea water and fresh ground water are miscible fluids, and a transition zone develops between them. Across this zone the density of the water varies gradually from that of fresh water to that of sea water.

Another case in which a transition zone develops as a result of hydrodynamic dispersion is that of introducing water by various artificial recharge methods (surface spreading techniques or injection through wells), one quality of water into an aquifer containing water of another quality. The different quality serves in this case as a tracer by means of which the fluids and their mixture are recognized. Of special interest in this respect are the cases of waste disposal, sewage or radioactive wastes, into aquifers (sometimes into deep formations).

Under certain conditions, the transition zone in these cases is narrow, say with respect to the dimensions of the injected water body or with respect to those of the aquifer, so that the abrupt interface approach may be implemented in order to investigate the movement of this transition zone.

Unsaturated flow is a term often used by hydrologists and soil scientists to describe the flow of water and water vapor through a soil in which the void space is partly filled with air. Essentially, this is a case of simultaneous flow of two immiscible fluids (water and air) except that often the assumption is made that the air is practically immobile. As the focus of attention of this chapter is on the flow of two liquids, excluding cases of gas or air flows, or flows with change of phase resulting from temperature or pressure variations, the case of unsaturated flow will not be treated here.

II. Interfacial Tension and Capillary Pressure

A brief review of some concepts that are essential to the understanding of phenomena of simultaneous flow of immiscible fluids in porous media is given here.

A. Porosity

The *porosity* of a porous medium is a *continuum* concept. According to this concept, the porous medium and the fluids filling its voids with their respective properties are replaced by fictitious *interpenetrating continua*, each of which occupies the entire flow domain. This is done by averaging point values (i.e., values at a microscopic level) of the various parameters over certain elementary volumes of the porous medium. The procedure is similar to that of defining the density of a fluid [5]. By the very definition of density, or viscosity, of a fluid, a passage is made from a molecular *microscopic* level to a *macroscopic description* of a continuum. Whenever a passage of this type is performed because of the difficulty of describing phenomena at the microscopic level in an exact manner, macroscopic coefficients are introduced. Actually, three levels of treating phenomena in a porous medium can be distinguished: a molecular one, a microscopic one (in which a point in the liquid continuum filling the void space is being considered—relating density, viscosity, and velocity to the fluid at that point—and a macroscopic one in which variables are referred to points in the porous medium. Coefficients are introduced at the passage from the molecular level to the microscopic one (e.g., viscosity of a fluid), and from the microscopic level to the macroscopic one. Porosity and permeability are examples of such coefficients, reflecting the hydroscientist's inability to describe, in a microscopic fashion, the solid matrix of a porous medium and its resistance to the flow. Actually, even if the microscopic level of phenomena could be described in an exact manner, there still remains the question of expressing these phenomena in measurable quantities. These, again, are macroscopic in nature, reflecting the overall (or average) behavior of the measured property within a certain volume in the vicinity of a mathematical point in the considered medium. Thus, in the continuum approach there exists a fictitious medium to which at every (physical) point (in the voids as well as in the solid matrix), kinematic and dynamic variables, or medium properties, which are continuous functions of the spatial coordinates of that point, can be assigned. The continuum approach also replaces the (unknown) equations describing microscopic phenomena by use of (macroscopic) partial differential equations into which the coefficients are introduced to accommodate the observed phenomena.

These considerations lead to a definition of a *representative elementary volume* (REV) of a porous medium such that

(1) It is large enough, i.e., contains a sufficiently large number of pores so as to enable a meaningful (measurable) statistical average when the actual medium is converted into a continuum described by continuous variables.

(2) It is small in comparison to the dimensions of the flow domain under consideration, so that it may be regarded as a *physical point* [5] or as a *material point*.

The size of the REV is determined by the concept of *porosity* in a way similar to that by which a liquid continuum is introduced through the concept of *density*. It is the smallest volume around a considered (mathematical or geometrical) point such that adding to it or subtracting from it one or several pores has a negligible effect on the value of the (volumetric) porosity n.

Let P be a mathematical point in the medium, enclosed by a volume ΔU for which P is the centroid. The ratio $\Delta U_v/\Delta U$, where ΔU_v is the void volume within ΔU, is the average porosity n of ΔU

$$n(\Delta U) = \Delta U_v/\Delta U \tag{1}$$

Computing this ratio for a sequence of gradually decreasing volumes $\Delta U_1 > \Delta U_2 > \Delta U_3$, etc., gives a sequence of values $n_i(\Delta U_i) = (\Delta U_v)_i/\Delta U_i$, $i = 1, 2, 3 \ldots$. For large values of ΔU in an inhomogeneous medium, $n(\Delta U)$ may vary appreciably. However, below a certain value of ΔU, the fluctuations will tend to decay, with small fluctuations resulting from the random nature of the arrangement and size of pores in the vicinity of P. By reducing ΔU below some value ΔU_0, large fluctuations are observed in n_i. This happens when the dimensions of ΔU_i become of the order of magnitude of the single pore. As $\Delta U \to 0$, n will become 0 or 1 depending on whether P is within the solid or the liquid phase (Fig. 1).

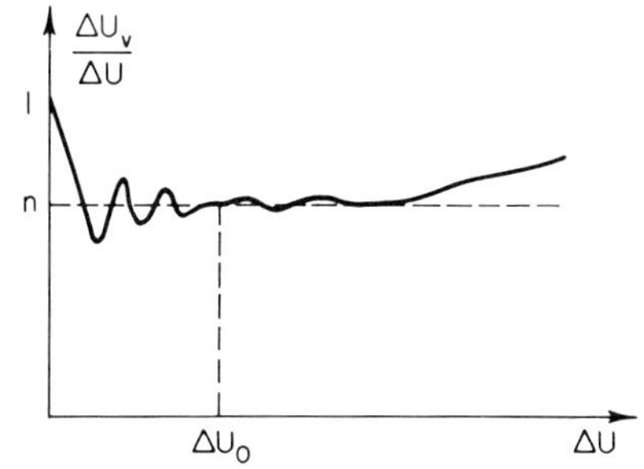

FIG. 1. Definition of porosity and REV.

The volumetric (or bulk) porosity n at P is defined by

$$n(P) = \lim_{\Delta U_i \to \Delta U_0} n\{\Delta U_i(P)\} = \lim_{\Delta U_i \to \Delta U_0} \{(\Delta U_v)_i/\Delta U_i\} \tag{2}$$

The volume ΔU_0 (Fig. 1) is the REV or the physical point of P. Continuity of n requires: $n(P') = \lim_{P \to P'} n(P)$.

It can be shown [6] in defining an *areal porosity* n' in a similar manner, that its average over an REV is equal to the (volumetric) porosity n. Hence, the average velocity vector $\mathbf{V}$ and specific discharge vector $\mathbf{q}$ (=volume of fluid per unit area of bulk medium per unit time) can be shown [6] to be related to each other by

$$\mathbf{V} = n\mathbf{q} \tag{3}$$

From here on, except when otherwise stated, variables and coefficients considered at a point within a porous medium will have the meaning of average values taken over an REV for which this point is a centroid.

B. Saturation

The void space of the porous medium can be filled by two or more immiscible fluids (liquids or gases). The *saturation* at a point within a porous medium with respect to a particular fluid is defined as the fraction of the void volume of the porous medium occupied by that particular fluid around the considered point

$$S_i = \frac{\text{volume of fluid } i \text{ within an REV}}{\text{total volume of voids within an REV}} \tag{4}$$

For two or more fluids,

$$\sum_{(i)} S_i = 1 \tag{5}$$

C. Interfacial Tension

When a liquid is in contact with another substance (another liquid immiscible with the first, a gas, or a solid), there is a free *interfacial energy* present between them. This reflects the amount of work that has to be performed in order to separate them. The interfacial energy arises from the difference between the inward attraction of the molecules in the interior of each phase and those at the surface of contact. Since a surface possessing free energy contracts if it can do so, the free interfacial energy manifests itself as *interfacial tension* [*1*]. Thus, the interfacial tension σ_{ik}, which is a constant for any pair of substances i and k, is defined as the amount of work that has to be performed in order to separate a unit area of substance i from substance k (units: erg/cm^2). The interfacial tension σ_i between a substance i and a vacuum is called *surface tension*. The interfacial tension σ_{ik} between substance i and k and the corresponding surface tensions are related to each other by *Dupre's formula*:

$$W_{ik} = \sigma_i + \sigma_k - \sigma_{ik} \tag{6}$$

where W_{ik} is the amount of work required to separate a unit area of substance i from substance k.

Figure 2*a* shows two immiscible liquids and a third fluid (liquid or gas) in contact with each other. Equilibrium requires that

$$\sigma_{AG} = \sigma_{AB} \cos \theta_{AB} + \sigma_{GB} \cos \theta_{GB} \tag{7}$$

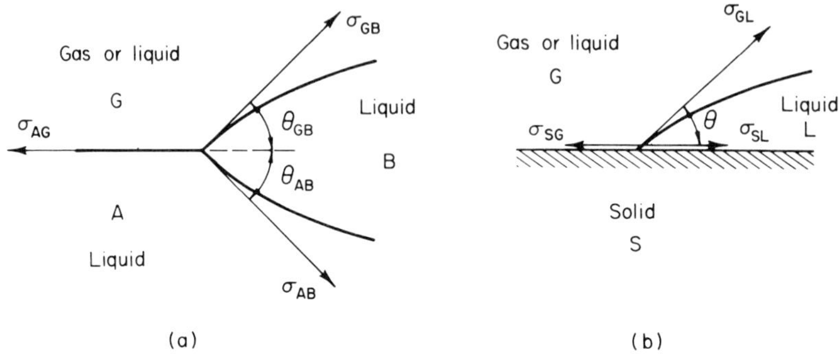

FIG. 2. Interfacial tensions. (*a*) immiscible liquids, (*b*) liquid-gas-solid.

which can be satisfied if $\sigma_{AG} < (\sigma_{AB} + \sigma_{GB})$, and a lens of liquid *B* will be formed. If $\sigma_{AG} > (\sigma_{AB} + \sigma_{GB})$, equilibrium is not possible, and liquid *B* will spread out between *A* and *G*.

Figure 2*b* shows two immiscible liquids (or a liquid and a gas) in contact with a solid surface. In this figure, θ denotes the angle between the *interface* and the surface *SL*. By convention, $0 < \theta < 180$ is measured through the denser fluid. Equilibrium requires that

$$\sigma_{GL} \cos \theta = \sigma_{SG} - \sigma_{SL}, \qquad \cos \theta = \frac{\sigma_{SG} - \sigma_{SL}}{\sigma_{GL}} \tag{8}$$

Equation (8), called *Young's equation*, states that $\cos \theta$ is defined by the ratio of the energy released in forming a unit area of interface between solid and liquid to the energy required for forming a unit area of fluid *G* and liquid *L*. Sometimes a factor is introduced to account for roughness in the solid's surface. Bikerman [7] questions Eq. (8) as this equation does not consider the force components normal to the surface.

From Eq. (8) it follows that no equilibrium is possible if $(\sigma_{SG} - \sigma_{SL})/\sigma_{GL} > 1$. In this case, liquid *L* will spread indefinitely over the solid. This leads to the concept of *wettability* of a solid by a liquid.

The angle θ in Eq. (8) is called the *contact angle*. The product $\sigma_{GL} \cos \theta$ is called *adhesion tension*; it determines which fluid will *preferentially wet* the solid, i.e., adhere to it and spread over it.

When $\theta < 90°$, liquid *L* (Fig. 2*b*) is said to wet the solid and is called the *wetting fluid*. When $\theta > 90°$, fluid *G* is called a *nonwetting fluid*. A zero adhesion tension indicates that both fluids have an equal affinity for the surface. In any system similar to that shown in Fig. 2*b*, it is possible to have either a fluid-*L*-wet or a fluid-*G*-wet surface, depending on the chemical composition of the fluids and the solid. Wettability has, therefore, a relative meaning only.

Interfacial tension, and hence also contact angle and wettability, may not be the same when a fluid-fluid interface (e.g., an air-liquid interface) is advancing or receding on a solid surface because of the *phenomenon of hysteresis*.

In an oil reservoir the fluids are usually oil and water, and we speak of an *oil-wet* or a *water-wet* rock. Most reservoirs are assumed water wet. Some rocks, however, seem to be partially wet to more than one fluid. Figure 3 shows the fluid distribution in an oil-wet and a water-wet rock.

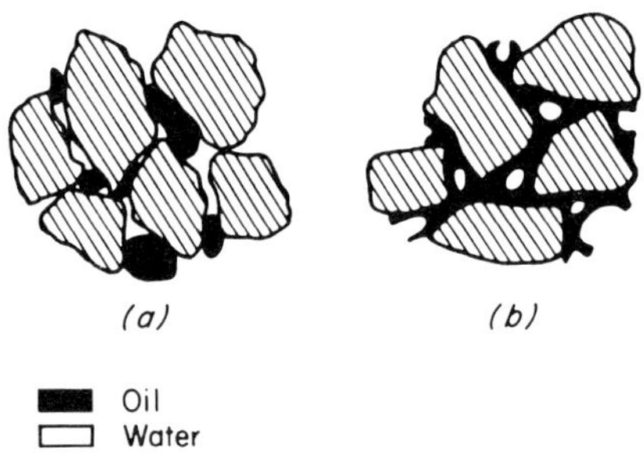

FIG. 3. Fluid distribution in the pore space as a function of fluid wettability [8]. (*a*) water wet sand, (*b*) oil wet sand.

With this concept of wettability, three types of fluid saturation between the limits 0 and 100 per cent [8] can be distinguished. Figures 4*a*, *b*, and *c* show a water-wet sand. Figures 4*d*, *e*, and *f* show an oil-wet sand. At very low water saturation (pendular saturation to water in Fig. 4*a*), water forms rings called *pendular rings* around the grain contact points. At this low water saturation the rings are isolated and do not form a continuous water phase except for a very thin film of water of nearly molecular thickness on the grains' surfaces. Practically no pressure can be transmitted from one ring

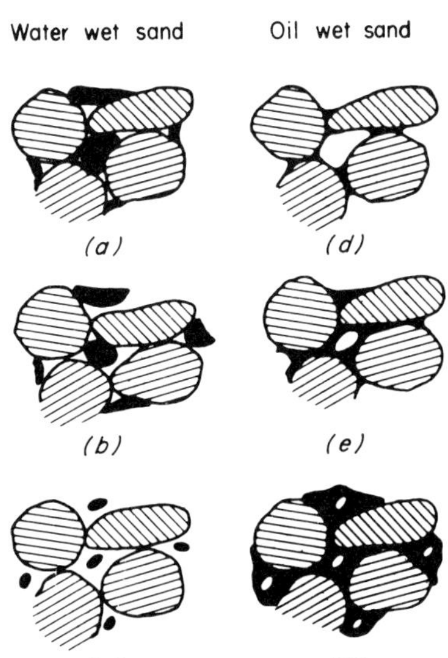

FIG. 4. Possible fluid saturation states. Saturations: (*a*) pendular to water, funicular to oil; (*b*) funicular to both oil and water; (*c*) funicular to water, insular to oil; (*d*) pendular to oil, funicular to water; (*e*) funicular to both oil and water; (*f*) funicular to oil, insular to water.

to the next within the water phase. As the wetting phase saturation increases, the pendular rings expand until a continuous wetting fluid phase is formed. The saturation at which this occurs is called *equilibrium saturation to the wetting phase*. Above this critical saturation, the saturation is called *funicular*, and flow of the wetting phase is possible. As the saturation of the wetting fluid increases, a situation develops in which the nonwetting fluid is no more a continuous phase; it breaks into individual droplets lodged in the larger pores (Fig. 4*c*). The nonwetting phase is then said to be in a state of *insular saturation*. A globule of this type can move only if a pressure difference sufficient to squeeze it through capillary restrictions is applied across it in the wetting phase. Similar considerations may be applied to the oil-wet sand (Figs. 4*d*, *e*, and *f*).

D. Capillary Pressure

When two immiscible fluids are in contact in the interstices of a porous medium, a discontinuity in pressure exists across the interface separating them. Its magnitude depends on the interface curvature at that point. Here, "point" is the microscopic point inside the void space. The difference in pressure (p_c), called *capillary pressure*

$$p_c = p_{nw} - p_w \tag{9}$$

where p_{nw} is the pressure in the nonwetting phase and p_w is the pressure in the wetting phase, can be determined by considering equilibrium on an elementary area around a point on the interface (Fig. 5). The local interface

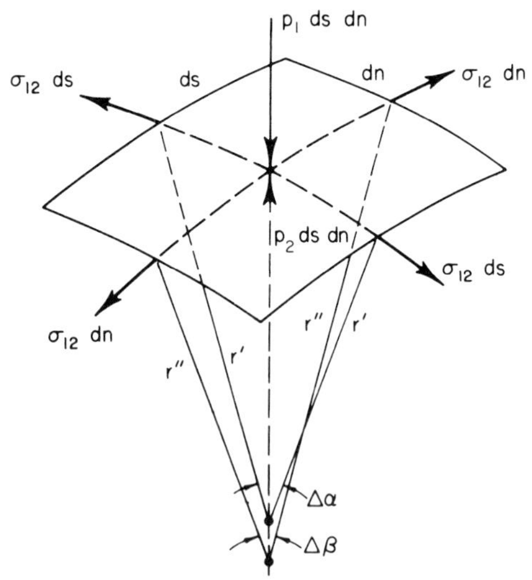

FIG. 5. Equilibrium at a curved interface between two immiscible fluids.

within a pore is curved and has two principal radii of curvature r' and r'' in two orthogonal planes. By considering the change in direction of the forces acting on opposite sides, we obtain

$$\Delta p \equiv p_c = p_2 - p_1 = \sigma_{12}(1/r' + 1/r'') = 2\sigma_{12}/r^* \tag{10}$$

where r^* is the mean radius of curvature ($2/r^* = 1/r' + 1/r''$). Equation (10) is known as the *Laplace equation for capillary pressure.* The capillary pressure is thus a measure of the tendency of a porous medium to suck in the wetting fluid phase or to repel the nonwetting phase.

In an actual porous medium, all terms in Eq. (10) have the meaning of a statistical average taken over the void space in the vicinity of a considered point in the porous medium. Because of the dependency of p_c locally (i.e., within each pore) on σ_{12} and r^* (which is of the order of magnitude of the pore size), it depends on the geometry of the void space (e.g., in the form of pore size distribution), on the nature of the solids and the liquids (e.g., in terms of the contact angle θ), and on the degree of saturation (say S_w of the wetting fluid that determines the volume of fluid accumulating as pendular rings at the points of contact between grains). In natural porous media, the geometry of the void space is extremely complex and cannot be defined analytically. Hence, the relationship $p_c = p_c(S_w)$ cannot be derived analytically (as can be done for simplified porous medium models, e.g., capillary tubes, spheres of constant radius, or a bundle of parallel circular rods). Instead, laboratory experiments have to be performed in order to determine this relationship. Figure 6 shows typical capillary pressure–wetting fluid saturation curves.

Several empirical and semiempirical expressions are available in the literature which attempt to relate the capillary pressure to medium and fluid properties and to S_w. Most of them are based on one or another idealized model of the porous medium. For example, in a capillary tube of radius r,

$$p_c = (2\sigma/r)\cos\theta \tag{11}$$

For other shapes of the cross-section, $2/r$ is replaced by $1/r'$ where r' is some equivalent radius equal to the ratio of volume to surface of the capillar. It is sometimes also called the *hydraulic radius of the capillary.* Leverett [9] suggested a semiempirical approach (using dimensional analysis), showing that the function

$$J = J(S_w) = (p_c/\sigma)\sqrt{k/n}; \quad p_c = p_c(S_w) \tag{12}$$

called the *Leverett function*, reduces to a common curve when plotted for several unconsolidated sands (Fig. 7). In Eq. (12), k is the medium's permeability, n is its porosity and the ratio $(k/n)^{1/2}$ may be interpreted as being proportional to the mean pore diameter.

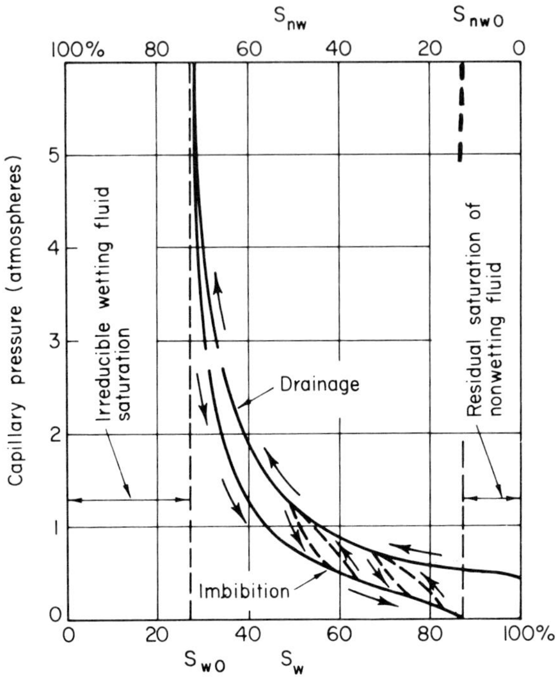

FIG. 6. Typical capillary pressure—wetting fluid saturation curves illustrating hysteresis.

Some authors suggest

$$J(S_w) = \frac{p_c}{\sigma \cos \theta} \left(\frac{k}{n}\right)^{1/2} \qquad \text{or} \quad J(S_w) = \frac{p_c}{\sigma f(\theta)} \left(\frac{k}{n}\right)^{1/2} \tag{13}$$

Brooks and Corey [*10*] show that when the experimental data $p_c = p_c(S_e)$, where $S_e = (S_w - S_{wo})/(1 - S_{wo})$ is the *effective*, or *reduced*, *saturation*, and S_{wo} is the irreducible wetting fluid saturation are drawn on a log-log paper, a straight line $S_e = (p_b/p_c)^\lambda$, $p_c \to p_b$ is obtained except for S_e close to unity. They suggest that the negative slope of this curve λ (*pore size distribution*) and the intercept p_b (= value of p_c obtained by extending the straight line to $S_e = 1$) be used as constants characterizing the medium. This approach is extremely useful, e.g., when choosing sand for a sand-box model, as the entire function $p_c = p_c(S_e)$ is represented by two constants.

Capillary pressure is subject to *hysteresis*, as the angle of contact θ is a function of the directional tendency to displacement; θ may have different values if equilibrium is approached by advancing or receding over a surface. For example, an advancing angle of contact occurs when water has a tendency to advance through oil, and a receding angle of contact when oil has a

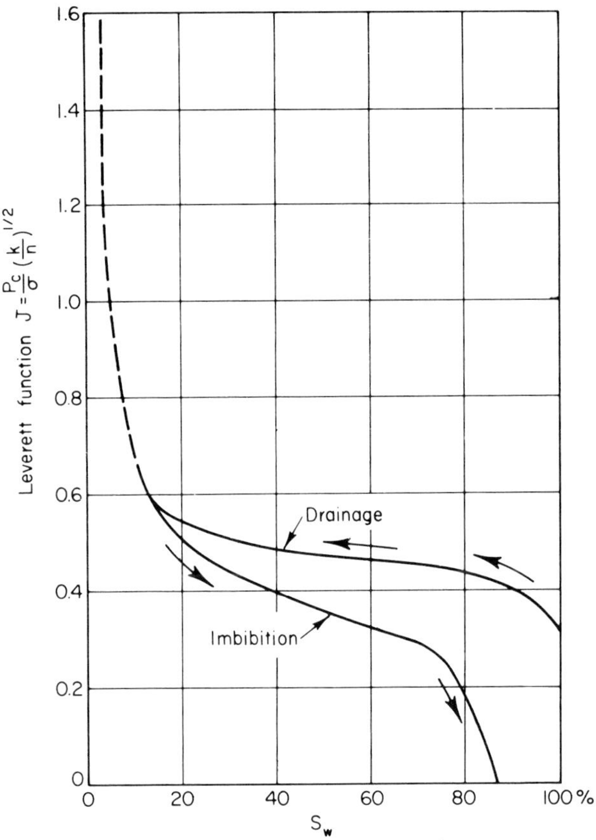

FIG. 7. Typical Leverett functions [9].

tendency to advance through water. Figure 8 shows this phenomenon for an air-water interface. As the interface advances along a plane surface, its curvature (reciprocal of radius of curvature) decreases; when the interface recedes,

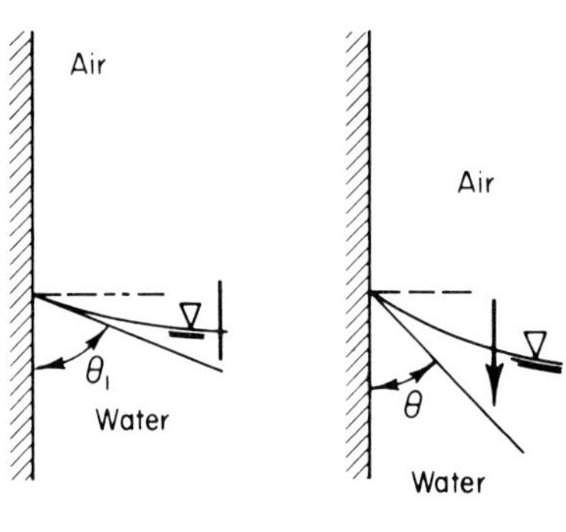

FIG. 8. Hysteresis in contact angle ($\theta_2 < \theta_1$).

its curvature increases, thus counteracting the drainage. Another reason for hysteresis is the geometry of the void space with many bottlenecks. (The *ink-bottle effect*.) The Leverett function given by Eq. (12) is also subject to hysteresis.

E. Drainage and Imbibition

Because of the *capillary hysteresis* phenomenon described before, different capillary pressure curves $p_c = p_c(S_w)$ may be obtained, depending on whether a sample is saturated initially with a wetting or a nonwetting fluid. In each case, the fluid that initially saturates the sample is slowly displaced by the other fluid.

When a sample is initially saturated with a wetting fluid, the process is called *drainage*, and the curve $p_c = p_c(S_w)$ is called the *drainage curve*. The process by which a wetting fluid displaces, by capillary forces alone, a nonwetting fluid which initially saturates a porous medium sample, is called *imbibition*. For example, if a sample is completely saturated with a nonwetting fluid, and some wetting fluid is introduced on its surface (or the sample is immersed in a wetting fluid), the wetting fluid will tend to flow *spontaneously* in along the solid walls of the pores, thus displacing the nonwetting fluid. The wetting fluid is said to displace the nonwetting one by imbibition. Imbibition is thus displacement by capillary forces only. Equilibrium is reached when the wetting fluid has accumulated in those pores which permit the greatest curvatures of the fluid-fluid interfaces, i.e., in the smallest pores; under such conditions, capillary forces equal those of gravity.

The curve describing the relationship $p_c = p_c(S_w)$ during imbibition (i.e., rise in wetting fluid saturation) is called the *imbibition curve*. Figure 6 shows a typical capillary pressure-wetting fluid saturation relationship including the effect of hysteresis (e.g., kerosene and water in a sandstone).

Following these curves, start from a sample that is completely saturated by a wetting fluid. As this fluid is slowly displaced, follow the drainage curve. Observe that a certain quantity of wetting fluid remains in the sample even at high pressures. The value of S_w at this point (denoted by S_{wo}) is called *irreducible saturation of the wetting fluid* (or *connate water saturation* in the case of water). Starting from a sample at this value of S_w and slowly displacing the nonwetting fluid by a wetting one, the imbibition curve is obtained. Observe that at zero capillary pressure there remains a certain amount of the nonwetting fluid; this is the *residual saturation* of the nonwetting fluid (denoted by S_{nwo}).

It is also possible to start the imbibition process from any point on the drainage curve, or to start the drainage process from any point on the imbibition curve (indicated by the dashed lines in Fig. 6). In this way the capillary pressure depends not only on the saturation at a certain instant, but also on the history of the particular sample under consideration. For a given capillary pressure, a higher saturation is obtained when the sample is being drained than when imbibition takes place.

When a sample is initially completely saturated by a wetting fluid ($S_w = 1$), displacement by a nonwetting fluid cannot take place unless the pressure in the latter is increased above a certain critical value that depends on the size of the pores. This pressure is called *nonwetting fluid entry value* (a threshold pressure or capillary pressure). In other words, a certain capillary pressure ($p_c = p_{nw} - p_w$, where p_{nw} is nonwetting pressure and p_w is wetting pressure) must be built up at the interface between the two fluids before drainage of the wetting fluid starts.

In most fluid flow problems, capillary hysteresis is not a serious problem, as the flow regime (drainage or imbibition) usually dictates which curve should be used.

F. Saturation Discontinuity at a Discontinuity in Medium

Let two porous media, 1 and 2, be in contact along a surface. A layer of coarse sand (1) overlying a stratum of tight consolidated sandstone (2) may serve as an example. The capillary pressure curves of the media are shown in Fig. 9. The two media are assumed to be water wet. Let a two-phase fluid mixture (e.g., oil and water or air and water) be present (under static or dynamic conditions) on both sides of the surface of contact. Given the wetting fluid saturation S_{w1} in medium 1, it is required to determine the saturation S_{w2} in medium 2. Since it was assumed that no fluid transfer takes place across the discontinuity, and the saturations are assumed to be funicular in both media, continuity of pressures across the boundary must exist. Hence, at the surface of contact,

$$p_{w1} = p_{w2}; \quad p_{nw1} = p_{nw2}; \quad p_{c1} = p_{c2}$$

The horizontal line in Fig. 9 connects the two equal values of p_c on both curves, and is used to determine S_{w2}. This means that a discontinuity in S_w must exist at a boundary of this kind.

The phenomenon just described occurs also at an interface between a liquid continuum and the porous medium when this interface serves as an external boundary of the flow domain. In this case, the phenomenon is known as *end effect* or *capillary end effect* [*8*].

III. Simultaneous Flow of Two Immiscible Liquids

A. The Basic Motion Equations

Many investigators (e.g., Chatenever and Calhoun [*11*]) conclude from experiments that when two immiscible fluids flow simultaneously through a porous medium each fluid establishes its own tortuous paths which form very

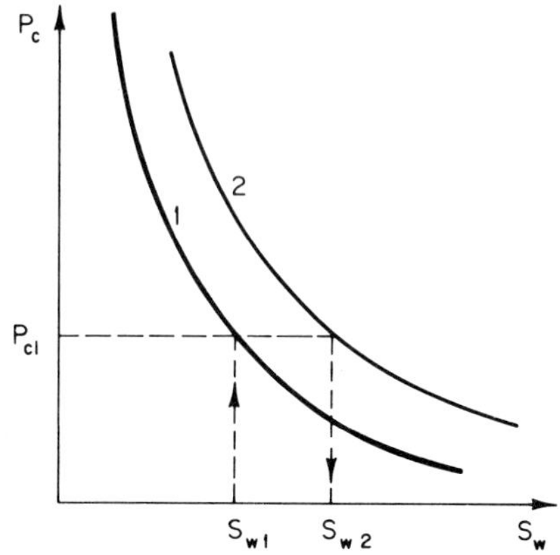

FIG. 9. Fluid saturation discontinuity between layers of different $p_c(S_w)$ characteristics.

stable channels. They assume that a unique set of channels corresponds to every degree of saturation. If a wetting fluid (S_w) and a nonwetting fluid (S_{nw}) are being considered, as S_{nw} is reduced, the channels of the nonwetting fluid tend to break down until only isolated regions of it remain (residual nonwetting fluid saturation). Similarly, as S_w decreases, the channels of the wetting fluid tend to break down and become discontinuous (irreducible wetting fluid saturation). When any of these fluids becomes discontinuous throughout the flow domain, no flow of it can take place.

With these concepts in mind, it seems natural to apply the concept of permeability established for the flow of a single phase fluid through a porous medium, modifying its value due to the presence of the second phase. Accordingly, it is possible to visualize an experiment of simultaneous *steady* flow of two fluids (1 and 2) through a horizontal porous medium column of constant cross-section A and finite length. Let pumps force constant discharge rates Q_1 of fluid 1 and Q_2 of fluid 2 through this column. Two pairs of manometers, placed a distance L apart along the column give the pressure drops Δp_1 in fluid 1 and Δp_2 in fluid 2. Once a steady flow has been established for both fluids, assume that Darcy's law, originally describing the flow of a single phase fluid completely saturating a porous medium, can be extended to describe the flow of each of the two immiscible fluids flowing simultaneously through the column

$$q^{(1)} = \frac{k^{(1)}}{\mu^{(1)}} \frac{\Delta p^{(1)}}{L}, \qquad q^{(2)} = \frac{k^{(2)}}{\mu^{(2)}} \frac{\Delta p^{(2)}}{L} \tag{14}$$

where q_i ($i = 1, 2$) is the specific discharge ($= Q_i/A$), and μ_i ($i = 1, 2$) is the dynamic viscosity of the ith fluid. In Eq. (14), k_1 and k_2 are called *effective permeabilities* of fluid 1 and of fluid 2, respectively. Obviously, these depend on the structure of the porous medium involved—specifically on the permeability k (dim: L^2) of the medium to a single phase fluid completely saturating it—and on the respective saturations (Sec. II, B). Often, the ratios

$$k_r^{(1)} = k^{(1)}/k, \qquad k_r^{(2)} = k^{(2)}/k \tag{15}$$

are used. They are called *relative permeabilities* of fluids 1 and 2, respectively.

Numerous experiments [*1*, p. 217] seem more or less to verify the validity of Eqs. (14), with the relative permeabilities depending on the nature of the porous medium, on the preferential wettability within the couple of flowing fluids, and on the saturations, but independent of the viscosities of the fluids and their specific discharges. However, it should be recalled that there is no way to derive k_{ri} ($i = 1, 2$) except through experiments such as the schematic one just described. In fact, the effective permeability is defined by Eqs. (14), or similar equations, which already presuppose the dependence of k_i or k and S_i alone. Additional comments on relative permeability are given in Sec. III, B.

Equations (14), as *working assumptions* describing the simultaneous flow of immiscible liquids with a sufficient accuracy for all practical purposes, can be generalized to include three-dimensional flows. For a single phase fluid (compressible or incompressible) completely saturating a porous medium, the motion equation (i.e., the equation relating the specific discharge vector $\mathbf{q}$ to the driving force) in an isotropic medium is

$$\mathbf{q} = -\frac{k}{\mu}(\text{grad } p - \rho\mathbf{g}) = -\frac{k}{\mu}(\text{grad } p + \rho g\, \mathbf{lz}) \tag{16}$$

where $\mathbf{g}$ is the gravity vector and $\mathbf{lz}$ is the unit vector in the $+z$ (vertically upward) direction, p is pressure, and ρ is the density of the fluid. In an isotropic medium, k is a scalar. In an anisotropic medium, the permeability k is a *symmetric second-order tensor*. According to *Einstein's summation convention* (or the *double index summation convention*), in any product of terms a suffix repeated twice (and only twice) is held to be summed over its three values 1, 2, and 3, and Eq. (16) becomes, for an anisotropic medium,

$$q_j = -\frac{k_{ij}}{\mu}\left(\frac{\partial p}{\partial x_i} + \rho g \frac{\partial z}{\partial x_i}\right); \qquad i, j = 1, 2, 3 \tag{17}$$

where the x_i's ($i = 1, 2, 3$) form the coordinate system. In the special case of a cartesian coordinate system x, y, z, where the directions of the coordinates also coincide with the principal directions of anisotropy, Eq. (17) may be written in the form

$$q_x = -\frac{k_x}{\mu}\frac{\partial p}{\partial x}, \qquad q_y = -\frac{k_y}{\mu}\frac{\partial p}{\partial y}, \qquad q_z = -\frac{k_z}{\mu}\left(\frac{\partial p}{\partial z} + \rho g\right) \tag{18}$$

where k_x, k_y, and k_z are the principal values of the anisotropic permeability.

For an incompressible fluid, Eq. (16) can be written

$$\mathbf{q} = -K \operatorname{grad} \varphi, \qquad \varphi = z + p/\rho g \tag{19}$$

where φ is the piezometric head (mechanical energy per unit weight of fluid) and $K = k\rho g/\mu$ is the hydraulic conductivity of the medium with respect to the particular fluid considered.

For a compressible fluid, where $\rho = \rho(p)$, Hubbert [12] introduced the potential

$$\varphi^* = \int_0^z dz + \int_{p_0}^p \frac{dp}{g\rho(p)} \tag{20}$$

where z_0 and p_0 are reference elevation and pressure, respectively. Equation (17) then becomes

$$\mathbf{q} = -\frac{k\rho g}{\mu} \nabla\varphi^* \tag{21}$$

Expressions for q_i ($i = 1, 2, 3$) may be derived from Eqs. (19) and (21) with K and k as second-rank symmetric tensors. A heuristic extension of these equations to the simultaneous flow of two immiscible fluids leads to the following equations for an isotropic porous medium

$$\begin{aligned} \mathbf{q}^{(1)} &= -\frac{k^{(1)}}{\mu^{(1)}}(\nabla p^{(1)} + \rho^{(1)} g \mathbf{1z}) = -k\frac{k_r^{(1)}}{\mu^{(1)}}(\nabla p^{(1)} + \rho^{(1)} g \mathbf{1z}) \\ \mathbf{q}^{(2)} &= -\frac{k^{(2)}}{\mu^{(2)}}(\nabla p^{(2)} + \rho^{(2)} g \mathbf{1z}) = -k\frac{k_r^{(2)}}{\mu^{(2)}}(\nabla p^{(2)} + \rho^{(2)} g \mathbf{1z}) \end{aligned} \tag{22}$$

or

$$\mathbf{q}^{(m)} = -\frac{k^{(m)}\rho^{(m)} g}{\mu^{(m)}} \nabla\varphi^{(m)} = -k\frac{k_r^{(m)}\rho^{(m)} g}{\mu^{(m)}} \nabla\varphi^{(m)}; \quad m = 1, 2 \tag{23}$$

where superscripts 1 and 2 denote the two fluids and $k^{(1)}$ and $k^{(2)}$ are the corresponding effective permeabilities. In this form, the equations were suggested by Wyckoff and Botset [13] and Muskat *et al.* [14]. For an anisotropic medium, Eqs. (22) become

$$\begin{aligned} q_i^{(1)} &= -\frac{k_{ij}^{(1)}}{\mu^{(1)}}\left(\frac{\partial p^{(1)}}{\partial x_j} + \rho^{(1)} g \frac{\partial z}{\partial x_j}\right) = -k_{ij}\frac{k_r^{(1)}}{\mu^{(1)}}\left(\frac{\partial p^{(1)}}{\partial x_j} + \rho^{(1)} g \frac{\partial z}{\partial x_j}\right) \\ q_i^{(2)} &= -\frac{k_{ij}^{(2)}}{\mu^{(2)}}\left(\frac{\partial p^{(2)}}{\partial x_j} + \rho^{(2)} g \frac{\partial z}{\partial x_j}\right) = -k_{ij}\frac{k_r^{(2)}}{\mu^{(2)}}\left(\frac{\partial p^{(2)}}{\partial x_j} + \rho^{(2)} g \frac{\partial z}{\partial x_j}\right) \end{aligned} \tag{24}$$

where the summation convention is implied. Or in a compact form:

$$q_i^{(m)} = -\frac{k_{ij}^{(m)}\rho^{(m)}g}{\mu^{(m)}}\frac{\partial\varphi^{(m)}}{\partial x_j} = -k_{ij}\frac{k_r^{(m)}\rho^{(m)}g}{\mu^{(m)}}\frac{\partial\varphi^{(m)}}{\partial x_j} \tag{25}$$

In Eq. (25) an assumption has been introduced that has yet to be verified—that $k_r^{(m)}$ is a scalar function of the saturation only, but independent of direction. For a compressible fluid, φ is replaced by φ^*. Recall that for a given (physical) point in the flow domain, p_1 and p_2 are pressures in the phase 1 and in the phase 2 fluids, respectively, each representing an average over the volume of fluid of that phase included in the REV for which the considered point is a centroid. From the discussion in Sec. II, D, it follows that at each such point $p^{(2)} - p^{(1)} = p_c$, assuming that p_c is the same as that of fluids at equilibrium. Like the relative permeability, the capillary pressure p_c depends on the saturation. Hence,

$$p^{(2)} - p^{(1)} = p_c(S^{(1)}) \tag{26}$$

B. Relative Permeability

Underlying the extension of the motion equation of a homogeneous fluid to the simultaneous flow of two or more fluids is the concept of *relative permeability*. At first sight, it seems natural to assume that when the flow of one of the fluids at a point is being considered, since part of the pore space in the vicinity of that point is occupied by another fluid, the permeability of the porous medium with respect to the considered fluid is reduced. This means that the relative permeability depends only on the saturation. Figure 10

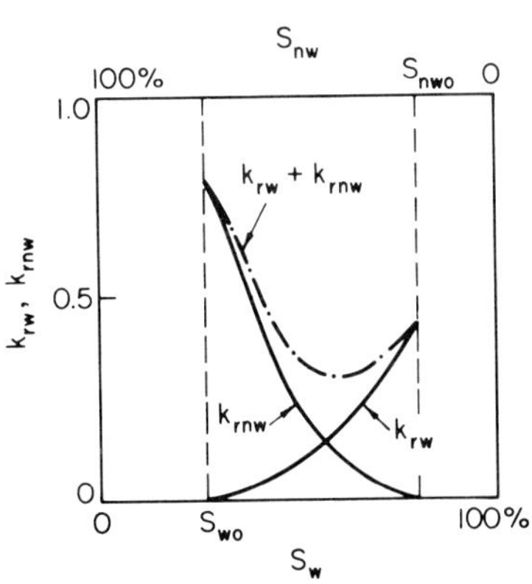

FIG. 10. Typical relative permeability curves (e.g., wetting fluid—water, nonwetting fluid—oil).

shows typical relative permeability curves for a pair of fluids: a wetting fluid (subscript *w*) and a nonwetting fluid (subscript *nw*). In each particular case, the exact shape of the curves depends on the porous medium and on the two fluids involved. In general, the curves have an S-shape. Obviously, both fluids are in motion only for a wetting fluid saturation which is above $S_w = S_{wo}$ (the irreducible wetting fluid saturation) and below $S_w = 1 - S_{nwo}$ where S_{nwo} is the residual nonwetting fluid saturation. At S_{nwo}, k_{rw} is usually much less than 1, whereas k_{rnw} at S_{wo} approaches the value of 1.

So far, there is no theoretical basis for this rather simplified picture. In fact, Eqs. (22) can be thought of as defining the relative permeability, assuming that the latter is independent of the pressure and the velocity of the fluid. Although numerous experiments (e.g., those listed in Ref. [*1*]) show that this assumption is indeed a good one for all practical purposes, many researchers have attempted to evaluate the influence of various fluid and flow parameters on the relative permeability.

One observation is that since one of the fluids preferentially wets the solid and adheres to its surfaces, the nonwetting fluid will be everywhere surrounded by the nonwetting fluid. This brings to mind a Poiseuille-type concentric flow in a tube which serves as a model of such a flow in a porous medium. This means that the picture of each fluid establishing its own channels of flow through the medium is rather questionable. Moreover, since wettability (Sec. II, C) is subject to hysteresis, it is evident that the relative permeability is also affected by this phenomenon, i.e., it depends on the manner in which the two fluids are distributed within the pore space. For the many possible equilibrium distributions of the two fluids within the pore space, there are many possible flows of the two fluids, and hence many possible pairs of values of relative permeability. This means that the relative permeability curves depend on the manner in which the experiment is performed. Figure 11 shows

FIG. 11. Effect of hysteresis on relative permeability.

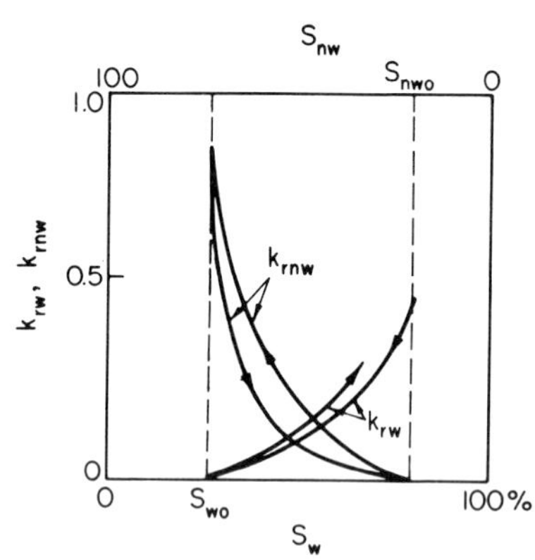

results of several such possible experiments with two liquids. The arrows indicate the direction in which the experiments were performed.

The effect of pressure gradients on the relative permeability has been investigated and reported by many authors. However, it seems that no definite conclusion has been reached. From available data, it seems [*2*] that for oil-water mixtures, the permeabilities, at least to the nonwetting phase, are higher at higher pressure gradients. Muskat [*2*] concludes that while physical considerations suggest that the equilibrium permeability and saturation values should not be independent entirely of the pressure gradient, neglecting such effects is warranted at least as a first approximation.

An effect of another factor on the relative permeability that has been studied and reported by many authors is the effect of the interference that one fluid exerts on the other within the pore space as a result of the difference in their viscosities. This effect is demonstrated by the fact that the sum $k_{rw} + k_{rnw}$ (indicated by the dashed line in Fig. 10) is less than unity (in other words, $(k_w + k_{nw}) < k$). Another example that seems to indicate that the viscosity ratio μ_w/μ_{nw} should affect k_{rw} and k_{rnw} is that the relative permeability to oil of a reservoir rock containing small amounts of connate water in pendular saturation (Fig. 4*a*) might be greater than unity, an observation that contradicts the assumption that k_{rw} or k_{rnw} in Eqs. (22)–(25) are independent of the fluids' viscosity. In the case of oil and water, Russell and Charles [*15*] state that the presence of a thin layer of water on the solid surfaces, in analogy to a thin layer of water on the internal surface of a pipe, acts as a "lubricant," thus reducing the resistance to the flow. All these phenomena indicate that a transfer of viscous forces takes place across the fluid-fluid interfaces within the pore space. Yuster [*16*], who also discussed the flow of fluid mixtures through porous media, suggested that a *finite velocity* must be taken as the boundary condition at the fluid-fluid interface. By considering a Poiseuille-type concentric flow in a circular tube (as a model representing the simultaneous flow of two immiscible fluids in a porous medium), he concluded that the viscosity ratio μ_w/μ_{nw} affects the specific discharge of the two fluids, and, hence, the extension of Darcy's law with permeabilities that are independent of viscosity ratio is not permitted. Another way of stating the same conclusion is that if Darcy's law is extended to include also the case of simultaneous flow of two fluids, the relative permeability defined by this law is a function of the viscosity ratio, in addition to being a function of the saturation. Milne–Thomson [*17*] and Hall [*18*] take the continuity of normal pressure and the viscous stress, provided surface tension is ignored, as the boundary conditions on the fluid-fluid interfaces. Scott and Rose [*19*] chose to assume that Yuster's capillary tube model overemphasized the magnitude of the ratio of fluid-fluid interfacial area to fluid-solid surface area. By replacing portions of the fluid-fluid interface by a thin impervious wall, they were able to demonstrate that there would be a corresponding decrease in the dependence of relative permeabilities upon the viscosity ratio.

Rose [20] started from the fact that one fluid passing over another contiguous immiscible fluid will always impart a motion to the second fluid. His model, representing the porous medium, has the geometry of a wide crack with parallel surfaces. The two phases *A* and *B* flowing parallel to the crack's axis are separated in part by a fluid-fluid interface, and in part by a thin impervious wall. Another model used by Rose [20] is a crack model with eddy pockets containing the phase *B* fluid circulating in a vortex-like motion. By solving for the flow in these models, caused by applying the driving force to one or both fluids, Rose [20] showed the magnitude of the "lubrication effect" and concluded that the volumetric throughput of a particular fluid that only partially fills the pore space is sometimes greater than its throughput when it completely fills the pore space. This condition of increased relative permeability is met when the saturation of the considered fluid approaches unity, as long as the contiguous fluid has a lower viscosity. He also showed the effect of viscosity ratio on the relative permeability resulting from a nonzero velocity at the fluid-fluid interfaces (Fig. 12). In general this seems to be a second-order effect, thus explaining why most experimental data fail to show this dependence.

From the foregoing discussion, it seems that in principle the equation of

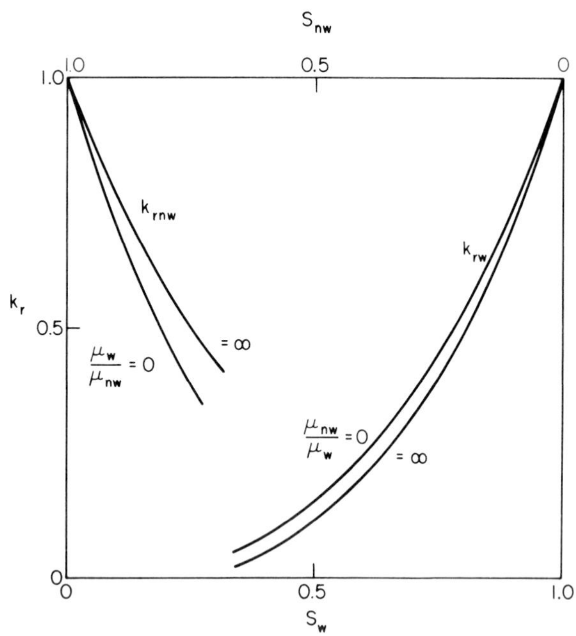

FIG. 12. The effect of viscosity ratio on relative permeability of a crack model with an eddy pocket [20].

motion cannot be extended formally to describe the simultaneous flow of two immiscible liquids, or, alternatively, that the concept of relative permeability defined by extending the equation of motion to this type of flow depends on several factors in addition to its dependence upon saturation. Nevertheless, available experimental evidence seems to indicate that this formal extension and the concept of a relative permeability that depends only on saturation is a good approximation for all practical purposes.

Many references on relative permeability are given in Refs. [*8*, p. 92; *1*, p. 217; *20* and *21*].

Scheidegger [*1*] and Pirson [*8*] list several methods for determining the effective and the relative permeability of a given sample. Some methods are based on steady flow of the two fluids at constant saturations. Other methods involve displacement of one fluid by another under unsteady flow conditions. In the latter case, the relative permeability is determined from the solution of the Buckley–Leverett equations (see Sec. III, D).

Instead of measuring directly the relative permeability by determining the two rates of flow (of the two fluids through a given core) and the saturation, some authors develop indirect methods based on the relationship between saturation and capillary pressure (Fig. 6) or on the pore size distribution. Such methods are based usually on some conceptual model of the porous medium for which the rate of flow can be determined in an exact manner. Burdine's [*22*] equation, for example, relates the relative permeabilities to the relationship $p_c = p_c(S_w)$, assuming that the nonwetting phase is oil and the wetting phase is water,

$$k_{rw} = S_e^{\,2} \frac{\int_o^{S_e} d(S_e')/(p_c^{\,2})}{\int_o^1 d(S_e')/(p_c^{\,2})}, \qquad k_{rnw} = (1 - S_e)^2 \frac{\int_{S_e}^1 d(S_e')/(p_c^{\,2})}{\int_o^1 d(S_e')/(p_c^{\,2})} \tag{27}$$

Since $p_c = p_c(S_e)$ has to be determined experimentally, the only advantage of using this approach is that it is sometimes easier and quicker to determine $p_c(S_w)$ than $k_{rw}(S_e)$.

Scheidegger [*1*, p. 240] reviews several other theoretical studies related to the concept of relative permeability.

C. Mass Conservation Equation

Several dependent variables appear in the equations of motion (22): $q_j^{(i)}(i = 1, 2; j = 1, 2, 3)$, $p^{(i)}(i = 1, 2)$, $S^{(i)}(i = 1, 2,$ in the relationships $k_r^{(i)} = k_r^{(i)}(S^{(i)}))$ and $\rho^{(i)}$ $(i = 1, 2)$. In the most general case (compressible fluids), there are here twelve dependent variables. Since Eqs. (22) include only six equations, it is obvious that six more relationships among the various variables are required in order to solve problems of simultaneous flow of fluids through porous media. These are

$$S^{(1)} + S^{(2)} = 1 \tag{28}$$

the equations of state of the fluids,

$$\rho^{(1)} = \rho^{(1)}(p^{(1)}), \qquad \rho^{(2)} = \rho^{(2)}(p^{(2)}) \tag{29}$$

the relationship between the capillary pressure p_c and the pressures $p^{(1)}$ and $p^{(2)}$,

$$p^{(2)} - p^{(1)} = p_c(S^{(1)}) \tag{30}$$

and two mass conservation equations, one for each fluid, to be considered in the following discussion.

An *extensive property* of a substance is one that depends on the amount (measured by mass) of the substance to which this property is referred. Volume, mass, and energy are examples of such properties. The property may be a scalar, a vector, or, in general a tensor of any rank. Let $G^\alpha(\mathbf{x}, t)$ denote an extensive property of a certain amount of substance α at time t contained within an arbitrary volume of space U centered at point $\mathbf{x}$ (vector symbol for coordinates x_1, x_2, x_3). With each extensive property, associate an *intensive property* that is independent of the amount of substance α present. The intensive property associated with G^α—denoted by γ^α—is the amount of that property per unit mass of substance α. Thus,

$$G^\alpha(\mathbf{x}, t) = \int_{(U)} \rho^\alpha(\xi, t)\gamma^\alpha(\xi, t)\, dU \tag{31}$$

where ξ (vector) denotes the coordinates of the centroid of the elementary volume dU. We may also introduce a volumetric density g^α associated with the same property, such that $g^\alpha = \rho^\alpha \gamma^\alpha$ is the amount of that property of the substance α per unit volume of the considered system (comprised in general of several substances).

Let the control volume approach, known in fluid mechanics as the Eulerian point of view, be used in order to study the conservation of the property G^α as the substance α moves within a specified domain. The control volume has a fixed shape and position in space, though the shape itself is arbitrary.

The conservation of the considered property of the substance α for the fixed control volume U can be expressed in the following way:

$$\int_{(U)} \frac{\partial g^\alpha}{\partial t}\, dU = -\int_{(S)} g^\alpha \mathbf{V}^\alpha \cdot \mathbf{dS} + \int_{(U)} I^\alpha\, dU \tag{32}$$

where I^α is a *source-sink function* expressing the temporal rate of production per unit volume (of the system) of the considered property within U through internal processes, and $\mathbf{V}^\alpha$ is the average velocity of the α-substance within the given flow domain. Equation (32) actually states that the rate of increase of G^α within U is equal to the sum of the rate at which the considered property crosses into U through the surface S of U, and the rate at which this property is produced within U, say by chemical reactions. In Eq. (32), the order of integration and differentiation (with respect to t) has been exchanged on the left-hand side, as U is constant. Applying Gauss theorem to the surface integral in Eq. (32), it is possible to rewrite the conservation statement as

$$\int_{(U)} \frac{\partial g^\alpha}{\partial t}\, dU = -\int_{(U)} \operatorname{div}(g^\alpha \mathbf{V}^\alpha)\, dU + \int_{(U)} I^\alpha\, dU \tag{33}$$

or,

$$\int_{(U)} \left[\frac{\partial g^\alpha}{\partial t} + \operatorname{div}(g^\alpha \mathbf{V}^\alpha) - I^\alpha\right] dU = 0 \tag{34}$$

Since the volume U is arbitrary, the integrand must vanish for Eq. (34) to be valid for all points within U. Hence,

$$\frac{\partial g^\alpha}{\partial t} + \operatorname{div}(g^\alpha \mathbf{V}^\alpha) = I^\alpha \tag{35}$$

is the conservation statement for the property G^α. In the passage from the integral form of Eq. (34) to the differential form of Eq. (35), it was assumed that the quantities in the former vary smoothly with space and time.

With these general considerations in mind, apply Eq. (35) to each of the two phases participating in the simultaneous flow. In this case ($\alpha = 1, 2$),

$g^\alpha = nS^\alpha \rho^\alpha$ = mass of fluid α per unit volume of porous medium; n = porosity of porous medium

$g^\alpha \mathbf{V}^\alpha = \rho^\alpha \mathbf{q}^\alpha$ = mass flux (mass per unit area per unit time) of fluid α

$I_\alpha = 0$, assuming that there are no sources or sinks in the flow domain.

We obtain, assuming n = constant,

$$n\,\frac{\partial(S^\alpha \rho^\alpha)}{\partial t} + \operatorname{div}(\rho^\alpha \mathbf{q}^\alpha) = 0; \quad \alpha = (1), (2) \tag{36}$$

These are the two continuity (or mass conservation) equations governing the simultaneous flow of the two fluids.

In principle there is here a complete mathematical statement of the problem of determining the simultaneous flow of two liquids in a given porous medium domain. The twelve dependent variables are $q_j^{(i)}$, $\rho^{(i)}$, $p^{(i)}$, $S^{(i)}$ ($i = 1, 2; j = 1, 2, 3$). The equations are the six equations in Eq. (24), and Eqs. (28)–(30) and (36). In some cases, $\mu^{(1)}$ and $\mu^{(2)}$ are also considered as dependent variables. Two equations of state $\mu^{(i)} = \mu^{(i)}(p^{(i)})$ are then added to enable a solution to be achieved. Obviously, appropriate initial and boundary conditions have to be specified with respect to each of the dependent variables in order to enable solution of specific cases. For example, initial conditions will include the values of $\rho^{(i)}$, $\mu^{(i)}$ (for compressible fluids), $p^{(i)}$ and $S^{(i)}$ at every point of the flow domain at some initial time taken as $t = 0$. Boundary conditions (including the case of a surface of discontinuity, in k) are the same as those encountered in the case of a single fluid, and need not be repeated here. The porous medium properties are n, $k_r^{(1)} = k_r^{(1)}(S^{(1)})$, $k_r^{(2)} = k_r^{(2)}(S^{(2)})$, and $p_c = p_c(S^{(1)})$. These have to be determined experimentally.

It is possible to insert the equations for $\mathbf{q}^{(i)}$ into Eq. (36), thus eliminating $\mathbf{q}^{(i)}$ and reducing the number of dependent variables. When the two liquids are incompressible, $\rho^{(1)} = \text{const.}$, $\rho^{(2)} = \text{const.}$, the problem reduces to

$$n\frac{\partial S^{(1)}}{\partial t} - \frac{\partial}{\partial x_i}\left[k_{ij}\frac{k_r^{(1)}}{\mu^{(1)}}\left(\frac{\partial p^{(1)}}{\partial x_j} + \rho^{(1)}g\frac{\partial z}{\partial x_j}\right)\right] = 0$$

$$n\frac{\partial S^{(2)}}{\partial t} - \frac{\partial}{\partial x_i}\left[k_{ij}\frac{k_r^{(2)}}{\mu^{(2)}}\left(\frac{\partial p^{(2)}}{\partial x_j} + \rho^{(2)}g\frac{\partial z}{\partial x_j}\right)\right] = 0 \tag{37}$$

$$S^{(1)} + S^{(2)} = 1, \qquad p^{(2)} - p^{(1)} = p_c(S^{(1)})$$

i.e., four equations for the unknowns $S^{(1)}$, $S^{(2)}$, $p^{(1)}$, $p^{(2)}$. For such a case and dealing with an incompressible fluid, it is also possible to introduce the piezometric heads

$$\varphi^{(1)} = z + p^{(1)}/\rho^{(1)}g, \qquad \varphi^{(2)} = z + p^{(2)}/\rho^{(2)}g \tag{38}$$

A special case of simultaneous flow of two immiscible fluids is the so-called *unsaturated flow*, often encountered in soil science, in which the two fluids are air and water (see Stallman [23]). However, in unsaturated flow it is assumed that the air is immobile and (neglecting gravity) at a constant pressure equal to the atmospheric pressure. All the concepts discussed in Sec. II are valid also for unsaturated flow. The capillary pressure, also called *suction* or *tension* (negative of the pressure head), is the pressure difference between the pressure in the gaseous phase and in the water (in the same sense of averaging over an REV). The discussion in Sec. II, D, remains valid. The equation of motion takes the form

$$\mathbf{q} = -K(\theta)\,\nabla\varphi \quad \text{or} \quad \mathbf{q} = -K(S)\,\nabla\varphi; \quad \varphi = z + p/\gamma$$

where $\theta = nS$ is the water content (volume of water per unit volume of medium), S is saturation (volume of water per unit volume of voids), $K(\theta)$ or $K(S)$ is the effective hydraulic conductivity $= k(\theta)\gamma/\mu$ or $= k(S)\gamma/\mu$, and $k(\theta)$ or $k(S)$ is the effective permeability. This equation of motion is the same as Eq. (22) or Eq. (25) written for one fluid, with the effective permeability being a *function* of the saturation.

The continuity equation (36) is also valid (α representing the water phase). Since the water may be regarded as incompressible ($\rho =$ const),

$$\partial\theta/\partial t + \nabla \cdot \mathbf{q} = 0$$

or

$$\partial\theta/\partial t - \nabla \cdot [K(\theta)\, \nabla\varphi] = 0$$

In general, an analytic solution of Eqs. (37) is not possible owing to the nonlinearity of the continuity equations resulting from the dependence of the relative permeability on the saturation, and, hence, also on the pressure. Another difficulty arises from the fact that the medium properties are given, in general, in the form of curves summarizing laboratory experiments performed on samples. Therefore, in most cases solutions are obtained numerically by means of high-speed digital computers. Sometimes, sand-box models [*24*] are used.

A special case in which an analytical solution can be derived is given in the following section.

D. The Buckley–Leverett Case

Buckley and Leverett solved Eqs. (37) assuming that gravity, capillarity, and liquid compressibility may be neglected [*25*]. These are called the *Buckley–Leverett assumptions.* Neglecting capillarity and liquid compressibility only, Eqs. (22), (30), and (36) for one-dimensional flow in the vertical (z) direction can be written as [*3*]:

$$q_z^{(1)} \equiv q^{(1)} = -\frac{kk_r^{(1)}}{\mu^{(1)}}\left(\frac{\partial p}{\partial z} + \rho^{(1)}g\right), \qquad q_z^{(2)} \equiv q^{(2)} = -\frac{kk_r^{(2)}}{\mu^{(2)}}\left(\frac{\partial p}{\partial z} + \rho^{(2)}g\right)$$

$$S^{(1)} + S^{(2)} = 1 \tag{39}$$

$$n\frac{\partial S^{(1)}}{\partial t} + \frac{\partial q^{(1)}}{\partial z} = 0, \qquad n\frac{\partial S^{(2)}}{\partial t} + \frac{\partial q^{(2)}}{\partial z} = 0$$

By introducing new variables

$$q = q^{(1)} + q^{(2)}, \qquad r^{(1)} = q^{(1)}/q, \qquad r^{(2)} = q^{(2)}/q = 1 - r^{(1)}$$

Equations (39) can be rewritten

$$q = -k\left[\left(\frac{k_r^{(1)}}{\mu^{(1)}} + \frac{k_r^{(2)}}{\mu^{(2)}}\right)\frac{\partial p}{\partial z} + \left(\frac{k_r^{(1)}\rho^{(1)}}{\mu^{(1)}} + \frac{k_r^{(2)}\rho^{(2)}}{\mu^{(2)}}\right)g\right] \tag{40}$$

$$r^{(1)} = \frac{k}{q}\,\frac{(\mu^{(2)}/kk_r^{(2)})q - (\rho_1 - \rho_2)g}{\mu^{(1)}/k_r^{(1)} + \mu^{(2)}/k_r^{(2)}} \tag{41}$$

$$\partial q/\partial z = 0 \tag{42}$$

$$q\,\frac{\partial r^{(1)}}{\partial z} + n\,\frac{\partial S^{(1)}}{\partial t} = 0 \tag{43}$$

If gravity is neglected also (or the flow is horizontal), terms involving g will drop out. If the boundary condition at $z = 0$ is $q^{(1)} = \text{const}$ and $q^{(2)} = \text{const}$, it follows from Eq. (42) that q is a constant independent of z or t. From Eq. (44) it follows that $r^{(1)} = r^{(1)}(S^{(1)})$, as the relative permeabilities depend on S_1 only. Equation (43) can be rewritten

$$q\,\frac{dr^{(1)}}{dS^{(1)}}\,\frac{\partial S^{(1)}}{\partial z} + n\,\frac{\partial S^{(1)}}{\partial t} = 0 \tag{44}$$

In the more general case, the boundary condition at $z = 0$ is $q = q(t)$. Then, we obtain from Eq. (42) for $z > 0$

$$q = q(t)$$

i.e., independent of z. Equation (44) then becomes

$$q(t)\,\frac{dr^{(1)}}{dS^{(1)}}\,\frac{\partial S^{(1)}}{\partial z} + n\,\frac{\partial S^{(1)}}{\partial t} = 0, \qquad \frac{q(t)}{n}\,\frac{dr^{(1)}}{dS^{(1)}}\,\frac{\partial S^{(1)}}{\partial z} + \frac{\partial S^{(1)}}{\partial t} = 0 \tag{45}$$

This is a nonlinear, first-order partial differential equation for the single dependent variable $S^{(1)} = S^{(1)}(z, t)$. It can be treated by the *method of characteristics.*

The characteristics of this equation are curves described by

$$\frac{dz}{dt} = \frac{q(t)}{n}\,\frac{dr^{(1)}}{dS^{(1)}} \tag{46}$$

and

$$\frac{dS^{(1)}}{dt} = 0 \tag{47}$$

where

$$\frac{dS^{(1)}}{dt} = \frac{\partial S^{(1)}}{\partial t} + \frac{\partial S^{(1)}}{\partial z}\,\frac{dz}{dt} \tag{48}$$

For $q(t) =$ constant, the characteristic curves are straight lines of constant saturation.

From Eq. (46) it follows that any saturation advances at a velocity given by the tangent to the curve $r^{(i)} = r^{(i)}(S^{(1)})$ at the point corresponding to this saturation, multiplied by the fluid's velocity $q(t)/n$. For $q =$ constant, the velocity W depends only on $S^{(1)}$ and is independent of time.

Certain initial conditions yield a multiple-valued solution $S^{(1)} = S^{(1)}(z, t)$. Such multiple-valued solutions, introduced by neglecting the capillary pressure, can be avoided by introducing a discontinuity in the saturation at some distance from the entrance to the column. Welge [*26*] suggests a graphical method for the determination of this distance.

Marle [*3*] and Oroveanu [*4*] give a detailed discussion of this problem.

IV. Immiscible Displacement

A. The Abrupt Interface Approximation

1. *The Interface*

In many cases of practical interest, the transition zone between the two liquids is narrow relative to the dimensions of the regions occupied by each liquid alone. Under such conditions it can be assumed that each liquid is confined to a *separate portion* of the flow domain. An approximation of this type is often used in studying fluid injection programs for recovery of oil from an oil reservoir. It is also used in investigations concerning the interface between fresh water and salt water in coastal aquifers, whenever the transition zone between the two miscible liquids is relatively narrow.

With this approximation in mind, let R be some porous medium region occupied by two immiscible, incompressible liquids 1 and 2 (Fig. 13), and let S_1 and S_2 denote the portions of the external boundary of R in contact

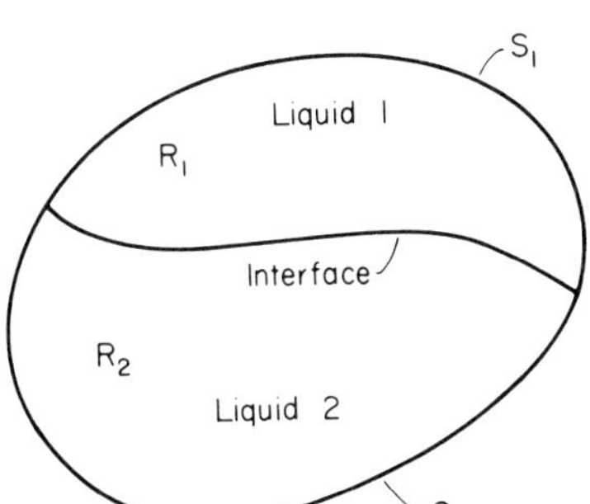

FIG. 13. An abrupt interface separating two liquids.

with liquids 1 and 2, respectively. A *piezometric head* (energy per unit weight) is defined within each subregion as

$$\begin{aligned} \varphi_1 &= \varphi_1(x, y, z, t) = z + p/\gamma_1 \qquad \text{in} \quad R_1 \\ \varphi_2 &= \varphi_2(x, y, z, t) = z + p/\gamma_2 \qquad \text{in} \quad R_2 \end{aligned} \tag{49}$$

where z is the elevation of the considered point above some datum level, p is pressure, and γ is the specific weight.

To solve the problem of flow of these liquids means to determine the piezometric heads φ_1 in R_1 and φ_2 in R_2 (Fig. 13) such that φ_1 and φ_2 each satisfies (i) an appropriate continuity equation in its respective subregion, (ii) initial conditions (in unsteady flow) in its respective subregion, and (iii) boundary conditions on its respective portion of the external boundary. In addition, φ_1 and φ_2 have to satisfy (simultaneously) boundary conditions on their common boundary—*the interface*. For example, if the two liquids are incompressible and the medium is homogeneous, isotropic, and non-deformable, the continuity equation to be satisfied by each liquid is Laplace's equation,

$$\begin{aligned} \partial^2\varphi_1/\partial x^2 + \partial^2\varphi_1/\partial y^2 + \partial^2\varphi_1/\partial z^2 &= 0 \qquad \text{in} \quad R_1 \\ \partial^2\varphi_2/\partial x^2 + \partial^2\varphi_2/\partial y^2 + \partial^2\varphi_2/\partial z^2 &= 0 \qquad \text{in} \quad R_2 \end{aligned} \tag{50}$$

The boundary conditions on S_1 and S_2 may be any of the ones encountered in the flow of a single fluid:

1. Boundary condition of the first type, i.e., of prescribed piezometric head, with an equipotential surface, $\varphi =$ constant, as a special case.
2. Boundary condition of the second type, i.e., a prescribed flux (e.g. by prescribing $\partial\varphi/\partial n$) along the boundary, with an impermeable boundary $\partial\varphi/\partial n = 0$ as a special case.

Sometimes, portions of $S_1 (\gamma_1 < \gamma_2)$ are a phreatic surface and a seepage face (neglecting the presence of a partially saturated region or a capillary fringe). The appropriate boundary conditions [*24*] should be prescribed on these portions.

The boundary conditions at the interface are derived in Sec. IV, B.

Assuming thus that the two liquids are completely separated by an abrupt interface, the velocity **V** of the interface at every point on it is related to the specific discharge **q** at that point by

$$\mathbf{V} = \mathbf{q}/n \tag{51}$$

where n is the porosity. Eq. (51) can be written twice, once for each of the two fluids. However, continuity considerations require that the component of the velocity normal to the interface be the same in both regions, or

$$V_n^{(1)} = q_n^{(1)}/n; \qquad V_n^{(2)} = q_n^{(2)}/n; \qquad V_n \equiv V_n^{(1)} = V_n^{(2)}; \qquad q_n^{(1)} = q_n^{(2)} \tag{52}$$

where subscript n indicates the direction of the normal to the interface. Sometimes, although the interface approximation may be employed, the separation is not a complete one. Instead, residual and irreducible saturations of immobile fluids exist. As an example, consider the case in a reservoir, of an interface between water and oil where the water is injected to displace the oil. The transition zone (where appreciable saturation gradients exist) is assumed to be narrow, so that on one side of the interface lie oil and immobile water at irreducible (or connate) water saturation S_{wo}, and on the other side lie water and immobile oil at irreducible oil saturation S_{nwo} (Fig. 10). Under these conditions, the interface velocity is

$$V_n = \frac{q_{n,\,\text{oil}}}{n(1 - S_{wo} - S_{nwo})} = \frac{q_{n,\,\text{water}}}{n(1 - S_{wo} - S_{nwo})} \tag{53}$$

$$q_{n,\,\text{oil}} = q_{n,\,\text{water}}$$

Here $n(1 - S_{wo} - S_{nwo})$ indicates the portion of a cross section through which flow can take place.

Using the summation convention, q_n can be expressed by

$$q_n = -\left(\frac{k_{ij}\rho g}{\mu}\frac{\partial\varphi}{\partial x_j}\right)n_i$$

where n_i are the components of the unit vector **In** in the direction normal to the interface.

2. *Piezometric Heads and Dynamic Equilibrium Conditions at a Stationary Interface* [27]

Let AB represent a small segment of a stationary interface completely separating two incompressible fluids in a two-dimensional flow in a vertical xy plane (Fig. 14*a*).

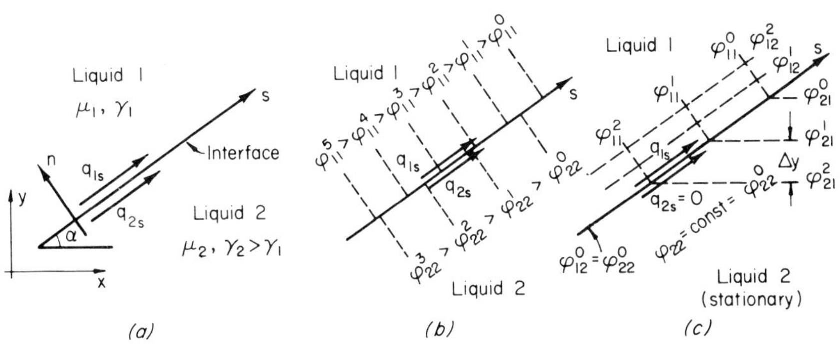

FIG. 14. Dynamic equilibrium conditions at a stationary interface.

The flow in each region is governed by Darcy's law

$$\mathbf{q}_i = -K_i \nabla\varphi_i = -\frac{k\gamma_i}{\mu_i}\nabla\left(y + \frac{p}{\gamma_i}\right) = -\frac{k}{\mu_i}(\nabla p + \gamma_i \nabla y) \tag{54}$$

where the fluids are denoted by $i = 1, 2$, and where K_i is the (isotropic) hydraulic conductivity to the ith liquid ($K_i = k\gamma_i/\mu_i$). Since each liquid is confined to its own subregion and the interface is stationary, the interface acts as an impervious boundary to each of the two subregions; only components of $\mathbf{q}_i (i = 1, 2)$ which are tangential to the interface are possible. If the interface is moving, continuity requires that the components of $\mathbf{q}_i (i = 1, 2)$ normal to it should be equal in magnitude. In addition, neglecting capillary pressure at the interface, continuity of pressure requires that the pressure be the same when a point on the interface is approached from both sides. Hence, the component of the pressure gradient tangential to the interface must also be the same on both sides of the interface.

With the nomenclature of Fig. 14*a*, the uniform flow in the direction s in both regions is given by

$$\begin{aligned} q_{1s} &= -\frac{k\gamma_1}{\mu_1}\frac{\partial\varphi_1}{\partial s} = -\frac{k}{\mu_1}\left(\frac{\partial p}{\partial s} + \gamma_1\frac{\partial y}{\partial s}\right) \\ q_{2s} &= -\frac{k\gamma_2}{\mu_2}\frac{\partial\varphi_2}{\partial s} = -\frac{k}{\mu_2}\left(\frac{\partial p}{\partial s} + \gamma_2\frac{\partial y}{\partial s}\right) \end{aligned} \tag{55}$$

By eliminating $\partial p/\partial s$ from both equations of Eq. (55),

$$\frac{\partial y}{\partial s} = \sin\alpha = \frac{q_{1s}\mu_1 - q_{2s}\mu_2}{k(\gamma_2 - \gamma_1)} \tag{56}$$

where $\alpha = \sin^{-1}(\partial y/\partial s)$ is the angle which the interface makes with the $+x$ axis. Following Hubbert [*27*], several conclusions may be drawn from Eq. (55).

a. For immobile liquid 2, $q_{2s} = 0$, and Eq. (55) becomes

$$\frac{\partial y}{\partial s} = \sin\alpha = \frac{q_{1s}\mu_1}{k(\gamma_2 - \gamma_1)}$$

or (57)

$$q_{1s} = \frac{k\gamma_1}{\mu_1}\frac{\gamma_2 - \gamma_1}{\gamma_1}\frac{\partial y}{\partial s} = K_1'\frac{\partial y}{\partial s}$$

with

$$K_1' = K_1\frac{\gamma_2 - \gamma_1}{\gamma_1} \qquad \text{and} \qquad K_1 = \frac{k\gamma_1}{\mu_1}$$

This means that the interfaces rise in the direction of the flow. As q_{1s} increases, α also increases. This occurs, for example, in a stationary coastal interface (Sec. VI) as the coast is approached.

For the sake of comparison consider the *phreatic surface* as an upper boundary of a fluid 1 flow domain in the vertical xy plane. On this surface $p = 0$ and $\varphi_1 = y$. The specific discharge component q_s tangent to this surface is given by

$$q_s = -K_1 \frac{\partial \varphi_1}{\partial s} = -K_1 \frac{\partial y}{\partial s} = -K_1 \sin \alpha \tag{58}$$

where α is the angle which the tangent to the phreatic surface makes with the $+x$ axis.

By comparing (57) with (58), observe the analogy that exists between a stationary phreatic surface and a stationary interface. The former drops in the direction of the flow, reaching at most an angle of $\alpha = \pi/2$; then $q_s = -K$, i.e., directed downward. The latter rises in the direction of the flow, reaching at most an angle of $\alpha = \pi/2$. Then, $q_s = K_1'$, directed vertically upward, or $(\partial\varphi_1/\partial s)_{\max} = -(\gamma_2 - \gamma_1)/\gamma_1$, which means that the component along the interface of the piezometric-head gradient cannot exceed a certain finite critical value. The analogy, shown here for a simple case of a stationary interface and free surface is, obviously, more general and exists also in the case of unsteady three-dimensional flow. We shall return to this problem in Sec. IV, B, where the analogy is shown between the boundary conditions on a phreatic surface and on an interface. This analogy is of interest, as sometimes it is more convenient to use it and study, e.g., by means of a Hele–Shaw analog [*24*] a problem involving an interface by investigating the corresponding phreatic surface problem.

b. If $q_{1s} = q_{2s} = q_s$, Eq. (56) becomes

$$\sin \alpha = q_s(\mu_1 - \mu_2)/k(\gamma_2 - \gamma_1) \tag{59}$$

If also $\mu_1 = \mu_2$, $\sin \alpha = 0$, i.e. the interface is horizontal. This is a quasi-stationary situation. We also have $\sin \alpha = 0$ when $q_s = 0$ (stationary horizontal interface) and when $q_{1s}\mu_1 = q_{2s}\mu_2$.

c. When $q_{1s} = 0$, $\sin \alpha = -q_{2s}\mu_2/k(\gamma_2 - \gamma_1)$. The interface will tilt downward in the direction of the flow. At most we may have $\alpha = -\pi/2$. Then $(\partial\varphi_2/\partial s) = -(\gamma_2 - \gamma_1)/\gamma_2$, which means that the component along the interface of the piezometric-head gradient cannot exceed a certain critical finite value.

d. When both liquids are in motion, α will attain some intermediate value, and the difference of the components along the interface of their piezometric-head gradient cannot exceed a certain finite critical value. If this critical value is exceeded (or when the critical values in cases *a* and *c* are exceeded), a stationary interface cannot exist. The interface will start moving until a new equilibrium is reached.

Let φ_{ij} denote the piezometric head (potential) in a region occupied by liquid i measured by the liquid level in an observation well filled with liquid j; then

$$\varphi_{ij} = y + p/\gamma_j; \quad i, j = 1, 2 \tag{60}$$

At any point within the flow domain the relationship between elevation and potentials has the form

$$y = [\gamma_2/(\gamma_2 - \gamma_1)]\varphi_{22} - [\gamma_1/(\gamma_2 - \gamma_1)]\varphi_{21} \tag{61}$$

or

$$y = [\gamma_2/(\gamma_2 - \gamma_1)]\varphi_{12} - [\gamma_1/(\gamma_2 - \gamma_1)]\varphi_{11} \tag{62}$$

Applying Eq. (62) to points along the interface and neglecting capillary pressure across the interface, the elevations along the interface are given by

$$y = [\gamma_2/(\gamma_2 - \gamma_1)]\varphi_{22} - [\gamma_1/(\gamma_2 - \gamma_1)]\varphi_{11} \tag{63}$$

If liquid 1 is flowing while liquid 2 (with $\gamma_2 > \gamma_1$) is stationary, the entire liquid 2 region is at a constant potential φ_{22}, say φ_{22}°. However, if in this region we measure potentials by means of liquid 1 (i.e. φ_{21}), equipotentials will be horizontal lines (Figs. 14*c* and 15). This follows from Eq. (61) with

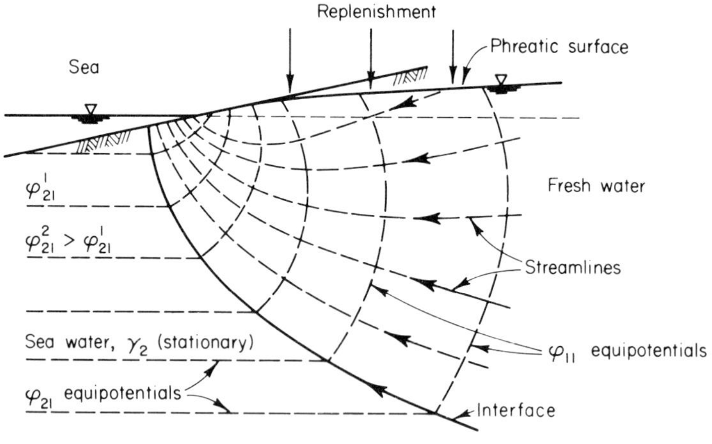

FIG. 15. Equipotentials near the coast.

$\varphi_{22} = \varphi_{22}^{\circ} = \text{const}$. Surfaces of $\varphi_{21} = \text{const}$ are the horizontal surfaces of $y = \text{const}$. The difference in elevation between two such adjacent surfaces is

$$(\Delta y)_{21} = -[\gamma_1/(\gamma_2 - \gamma_1)]\, \Delta\varphi_{21} \tag{64}$$

Surfaces of $\varphi_{12} = \text{const}$ are parallel to the interface (Fig. 14*c*).

Similar consideration can be applied to the case $q_{1s} = 0$, $q_{2s} \neq 0$, (Fig. 14*b*). In both cases, the potential gradients in the stationary liquid region will be vertical when expressed in terms of the density of the moving liquid.

Lusczynski [*28*], in connection with density variations across a coastal interface (Sec. VI), defined several types of head, also taking into account the presence of a transition zone:

(i). Point water head φ_{ip} at a point *i* (occupied by fluid *i*) in ground water of variable density is the water level (referred to a given datum level) in an observation well filled sufficiently with fluid *i* (of density ρ_i) so as to balance the pressure p_i at that point, or

$$\varphi_{ip} = z_i + p_i/\rho_i g \tag{65}$$

where *z* is the vertical coordinate.

(ii). Fresh water head φ_{if} is obtained when the observation well is filled with fresh water (of density ρ_f) to a level sufficient to balance the pressure p_i at point *i*, or,

$$\varphi_{if} = z_i + p_i/\rho_f g \tag{66}$$

(iii). Environmental water head φ_{in} is obtained when the observation well is filled with water of variable density, such that the variations of density are exactly the same as those encountered along a vertical line from point *i* to the phreatic surface. Lusczynski [*28*] related the three types of head to each other and determined the specific discharge components in terms of their gradients.

B. The Boundary Conditions on an Interface

In Sec. IV, A, 1, the solution of a problem involving an interface between two liquids was stated in terms of the two harmonic functions—φ_1 and φ_2. Next, we shall discuss the boundary conditions which φ_1 and φ_2 have to satisfy on the interface separating the two liquids.

The interface (stationary or moving) is a *material* (*or fluid*) *surface* that is always composed of the same fluid particles. This means that at each point on the interface, the component of the average velocity $\mathbf{V}$ normal to the interface in each of the regions is equal to the velocity of the interface. A portion of the interface in the vicinity of point *P* on it is shown in Fig. 16. The porous medium is assumed to be homogeneous with permeability *k* (or k_x, k_y, k_z in the case of anisotropy) and porosity *n*. It can be represented by an equation having the general form

$$F(x, y, z, t) = 0 \tag{67}$$

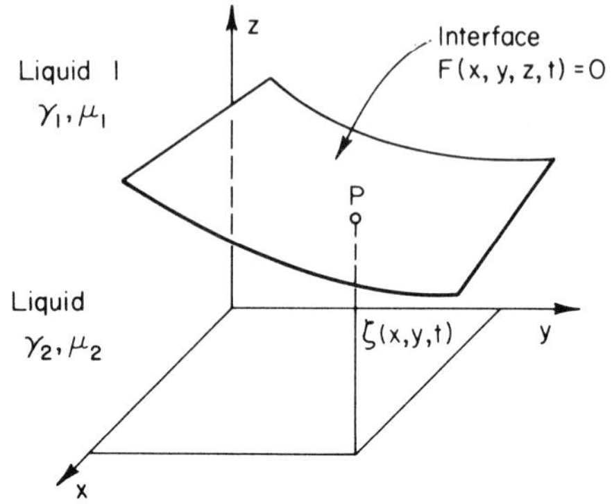

FIG. 16. An abrupt interface between immiscible liquids.

As the interface moves, a particle at a point (x, y, z) on it at time t will be at $(t + \Delta t)$ at $(x + dx, y + dy, z + dz)$, such that

$$F(x + dx, y + dy, z + dz, t + dt) = 0 \tag{68}$$

and dt is related to dx, dy, dz by

$$V_x = dx/dt = q_x/n, \qquad V_y = dy/dt = q_y/n, \qquad V_z = dz/dt = q_z/n \tag{69}$$

where subscripts x, y, and z denote components in these respective directions. As the interface moves,

$$\frac{dF}{dt} = \frac{\partial F}{\partial t} + \left(\frac{\partial F}{\partial x}\right)\left(\frac{dx}{dt}\right) + \left(\frac{\partial F}{\partial y}\right)\left(\frac{dy}{dt}\right) + \left(\frac{\partial F}{dz}\right)\left(\frac{dz}{dt}\right) = 0$$

or

$$dF/dt = \partial F/\partial t + (\mathbf{V} \cdot \nabla F) = \partial F/\partial t + (\mathbf{q} \cdot \nabla F)/n = 0 \tag{70}$$

where dF/dt (sometimes written as DF/Dt to emphasize that the derivative is taken following moving fluid particles) is the *substantial derivative*. Applying Eq. (70) to each of the two liquids,

$$\partial F/\partial t + (\mathbf{q}_1 \cdot \nabla F)/n = 0, \qquad \partial F/\partial t + (\mathbf{q}_2 \cdot \nabla F)/n = 0 \tag{71}$$

By relating the average velocity $\mathbf{V}$ to the specific discharge $\mathbf{q}$ through $\mathbf{V} = \mathbf{q}/n$, it is assumed that only one liquid is present on each side of the interface. In other cases, n should be replaced by $n(1 - S_{wo} - S_{nwo})$ as in Eq. (53), and $\mathbf{q}_1$ and $\mathbf{q}_2$ have to be expressed through the appropriate effective permeabilities: at S_{nwo} for the wetting fluid, and at S_{wo} for the nonwetting fluid.

With ζ denoting the elevation of a point on the interface (Fig. 16), $z = \zeta(x, y, t)$ on the interface, expression (67) becomes

$$F(x, y, z, t) \equiv z - \zeta(x, y, t) = 0 \tag{72}$$

By inserting Eq. (72) into Eq. (71), and expressing $\mathbf{q}_1$ and $\mathbf{q}_2$ by Eq. (54), we obtain the conditions

$$
\begin{aligned}
n\frac{\partial \zeta}{\partial t} - \frac{k\gamma_1}{\mu_1}\left(\frac{\partial \varphi_1}{\partial x}\frac{\partial \zeta}{\partial x} + \frac{\partial \varphi_1}{\partial y}\frac{\partial \zeta}{\partial y} - \frac{\partial \varphi_1}{\partial z}\right) &= 0 \\
n\frac{\partial \zeta}{\partial t} - \frac{k\gamma_2}{\mu_2}\left(\frac{\partial \varphi_2}{\partial x}\frac{\partial \zeta}{\partial x} + \frac{\partial \varphi_2}{\partial y}\frac{\partial \zeta}{\partial y} - \frac{\partial \varphi_2}{\partial z}\right) &= 0
\end{aligned}
\tag{73}
$$

which have to be satisfied on the interface, i.e., at $z = \zeta$.

A second condition to be satisfied on the interface is the dynamic one expressed by

$$p_2 - p_1 = p_c \tag{74}$$

By combining Eq. (74) and Eq. (49) written for $z = \zeta$,

$$\zeta = (\gamma_2/\Delta\gamma)\varphi_2 - (\gamma_1/\Delta\gamma)\varphi_1 - p_c/\Delta\gamma \tag{75}$$

Because of stability considerations, $\gamma_2 > \gamma_1$. Thus Eqs. (73) and (75) serve as boundary conditions on the abrupt interface. By inserting ζ from Eq. (75) into Eqs. (71) and (72), we obtain, for a homogeneous isotropic medium,

$$
\begin{aligned}
&\alpha_1 n\frac{\partial \Phi_1}{\partial t} - \alpha_2 n\frac{\partial \Phi_2}{\partial t} - \alpha_1\left[\left(\frac{\partial \Phi_1}{\partial x}\right)^2 + \left(\frac{\partial \Phi_1}{\partial y}\right)^2\right] \\
&\qquad + \alpha_2\left[\frac{\partial \Phi_1}{\partial x}\frac{\partial \Phi_2}{\partial x} + \frac{\partial \Phi_1}{\partial y}\frac{\partial \Phi_2}{\partial y}\right] - \frac{\partial \Phi_1}{\partial z} = 0 \qquad \text{on} \quad \zeta = z \\
&\alpha_1 n\frac{\partial \Phi_1}{\partial t} - \alpha_2 n\frac{\partial \Phi_2}{\partial t} + \alpha_2\left[\left(\frac{\partial \Phi_2}{\partial x}\right)^2 + \left(\frac{\partial \Phi_2}{\partial y}\right)^2\right] \\
&\qquad - \alpha_1\left[\frac{\partial \Phi_1}{\partial x}\frac{\partial \Phi_2}{\partial x} + \frac{\partial \Phi_1}{\partial y}\frac{\partial \Phi_2}{\partial y}\right] - \frac{\partial \Phi_2}{\partial z} = 0 \qquad \text{on} \quad \zeta = z
\end{aligned}
\tag{76}
$$

Since on the interface $1 + \alpha_1(\partial\Phi_1/\partial z) = \alpha_2(\partial\Phi_2/\partial z)$, Eq. (76) may be rewritten in the forms

$$
\begin{aligned}
&\alpha_1 n\frac{\partial \Phi_1}{\partial t} - \alpha_2 n\frac{\partial \Phi_2}{\partial t} - \alpha_1\left[\left(\frac{\partial \Phi_1}{\partial x}\right)^2 + \left(\frac{\partial \Phi_1}{\partial y}\right)^2 + \left(\frac{\partial \Phi_1}{\partial z}\right)^2\right] \\
&\qquad + \alpha_2\left[\frac{\partial \Phi_1}{\partial x}\frac{\partial \Phi_2}{\partial x} + \frac{\partial \Phi_1}{\partial y}\frac{\partial \Phi_2}{\partial y} + \frac{\partial \Phi_1}{\partial z}\frac{\partial \Phi_2}{\partial z}\right] - \frac{\partial \Phi_1}{\partial z} = 0 \qquad \text{on} \quad \zeta = z \\
&\alpha_1 n\frac{\partial \Phi_1}{\partial t} - \alpha_2 n\frac{\partial \Phi_2}{\partial t} + \alpha_2\left[\left(\frac{\partial \Phi_2}{\partial x}\right)^2 + \left(\frac{\partial \Phi_2}{\partial y}\right)^2 + \left(\frac{\partial \Phi_2}{\partial z}\right)^2\right] \\
&\qquad - \alpha_1\left[\frac{\partial \Phi_1}{\partial x}\frac{\partial \Phi_2}{\partial x} + \frac{\partial \Phi_1}{\partial y}\frac{\partial \Phi_2}{\partial y} + \frac{\partial \Phi_1}{\partial z}\frac{\partial \Phi_2}{\partial z}\right] - \frac{\partial \Phi_2}{\partial z} = 0
\end{aligned}
\tag{77}
$$

or in compact forms

$$
\begin{aligned}
&n\left(\alpha_1 \frac{\partial \Phi_1}{\partial t} - \alpha_2 \frac{\partial \Phi_2}{\partial t}\right) - \alpha_1 (\nabla \Phi_1)^2 + \alpha_2 (\nabla \Phi_1 \cdot \nabla \Phi_2) - \partial \Phi_1 / \partial z = 0 \\
&\qquad\qquad\qquad\qquad\qquad\qquad\qquad\qquad\qquad\qquad \text{on} \quad z = \zeta \\
&n\left(\alpha_1 \frac{\partial \Phi_1}{\partial t} - \alpha_2 \frac{\partial \Phi_2}{\partial t}\right) + \alpha_2 (\nabla \Phi_2)^2 - \alpha_1 (\nabla \Phi_1 \cdot \nabla \Phi_2) - \partial \Phi_2 / \partial z = 0
\end{aligned}
\tag{78}
$$

where $\alpha_i = \mu_i / k \Delta\gamma = \gamma_i / K_i \Delta\gamma$, $i = 1, 2$; K_i is the hydraulic conductivity with respect to the ith fluid, and $\Phi_i = K_i \varphi_i$ is the specific discharge potential of the ith fluid. We are permitted to introduce Φ because the medium is homogeneous and isotropic.

It is interesting to note that for $\Phi_1 \equiv 0$ and $\gamma_1 = 0$, $\gamma_2 = \gamma$, Eqs. (78) become the boundary condition on a phreatic surface (air-water interface).

Thus, our problem is to determine the two harmonic functions φ_1 and φ_2 (or Φ_1 and Φ_2) each in its respective subregion, satisfying Eqs. (50), boundary conditions on the external boundaries, and Eqs. (78) on the interface separating the two liquids. Once φ_1 and φ_2 are known, Eq. (75) may be used to determine the shape of the interface $\zeta = \zeta(x, y, t)$ or $F(x, y, z, t) = 0$. An exact solution of a moving interface by present available mathematical tools is most complicated because of the nonlinearity of Eq. (78) and because the latter is given on an *a priori* unknown surface $z = \zeta$. (Some approximate methods of solution are given in Sec. VI.) The shape of a stationary interface in two-dimensional flow in a vertical plane when one of the fluids is immobile can be derived by the hodograph method discussed in Sec. V.

If the porous medium is anisotropic, with x, y and z as principal directions, Eq. (73) becomes

$$
\left.
\begin{aligned}
&n \frac{\partial \zeta}{\partial t} - \frac{\gamma_1}{\mu_1}\left[k_x \frac{\partial \varphi_1}{\partial x}\frac{\partial \zeta}{\partial x} + k_y \frac{\partial \varphi_1}{\partial y}\frac{\partial \zeta}{\partial y} - k_z \frac{\partial \varphi_1}{\partial z}\right] = 0 \\
&n \frac{\partial \zeta}{\partial t} - \frac{\gamma_1}{\mu_1}\left[k_x \frac{\partial \varphi_2}{\partial x}\frac{\partial \zeta}{\partial x} + k_y \frac{\partial \varphi_2}{\partial y}\frac{\partial \zeta}{\partial y} - k_z \frac{\partial \varphi_2}{\partial z}\right] = 0
\end{aligned}
\right\} \quad \text{on} \quad \zeta = z \tag{79}
$$

and Eq. (77) becomes

$$
\begin{aligned}
&n \frac{\gamma_1}{\Delta\gamma}\frac{\partial \varphi_1}{\partial t} - n \frac{\gamma_2}{\Delta\gamma}\frac{\partial \varphi_2}{\partial t} - \frac{\gamma_1}{\Delta\gamma}\cdot\frac{\gamma_1}{\mu_1}\left[k_x\left(\frac{\partial \varphi_1}{\partial x}\right)^2 + k_y\left(\frac{\partial \varphi_1}{\partial y}\right)^2 + k_z\left(\frac{\partial \varphi_1}{\partial z}\right)^2\right] \\
&\qquad + \frac{\gamma_2}{\Delta\gamma}\frac{\gamma_1}{\mu_1}\left[k_x\left(\frac{\partial \varphi_1}{\partial x}\right)^2 + k_y\left(\frac{\partial \varphi_1}{\partial y}\right)^2 + k_z\left(\frac{\partial \varphi_1}{\partial z}\right)^2\right] - \frac{k_z \gamma_1}{\mu_1}\frac{\partial \varphi_1}{\partial z} = 0 \\
&n \frac{\gamma_1}{\Delta\gamma}\frac{\partial \varphi_1}{\partial t} - n \frac{\gamma_2}{\Delta\gamma}\frac{\partial \varphi_2}{\partial t} - \frac{\gamma_2}{\Delta\gamma}\cdot\frac{\gamma_2}{\mu_2}\left[k_x\left(\frac{\partial \varphi_2}{\partial x}\right)^2 + k_y\left(\frac{\partial \varphi_2}{\partial y}\right)^2 + k_z\left(\frac{\partial \varphi_2}{\partial z}\right)^2\right] \\
&\qquad - \frac{\gamma_1}{\Delta\gamma}\cdot\frac{\gamma_2}{\mu_2}\left[k_x\left(\frac{\partial \varphi_2}{\partial x}\right)^2 + k_y\left(\frac{\partial \varphi_2}{\partial y}\right)^2 + k_z\left(\frac{\partial \varphi_2}{\partial z}\right)^2\right] - \frac{k_z \gamma_2}{\mu_2}\frac{\partial \varphi_2}{\partial z} = 0 \\
&\qquad\qquad\qquad\qquad\qquad\qquad\qquad\qquad\qquad\qquad \text{on} \quad \zeta = z
\end{aligned}
\tag{80}
$$

C. Some Horizontal Interface Displacement Cases

Some simple cases of interface displacement will be considered now. In all cases the medium is assumed to be isotropic. If the medium is anisotropic, the set of transformations suggested by Bear and Dagan [*29*] should be applied to the solutions derived for isotropic media in order to obtain the corresponding solutions for anisotropic media.

1. *Linear Displacement*

Consider the case of flow of two incompressible fluids in the horizontal x direction in a homogeneous porous medium. The two fluids are separated by an interface in the form of a plane having a normal in the direction x. As this interface advances, let fluid 1 (wetting, water) displace fluid 2 (nonwetting, oil). At some instant t, the interface is at a distance ξ from the origin $x = 0$. Hence, the interface equation (67) becomes

$$F(x, y, z, t) \equiv x - \xi(t) = 0 \tag{81}$$

The problem may be formulated in terms of the pressure (since the flow is horizontal) as follows: Determine p_w and p_{nw} (or determine $\xi = \xi(t)$) such that for continuity,

$$\partial^2 p_w/\partial x^2 = 0; \quad 0 < x < \xi(t) \tag{82}$$

$$\partial^2 p_{nw}/\partial x^2 = 0; \quad \xi(t) < x < L \tag{83}$$

and on the interface,

$$p_w = p_{nw} \qquad \text{(neglecting } p_c\text{) on} \quad \xi = \xi(t) \tag{84}$$

$$\frac{k_{wo}}{\mu_w}\frac{\partial p_w}{\partial x} = \frac{k_{nwo}}{\mu_{nw}}\frac{\partial p_{nw}}{\partial x} = -q_x \qquad \text{on} \quad \xi = \xi(t) \tag{85}$$

where k_{wo} is the permeability to the wetting fluid (water) at residual nonwetting fluid (oil) saturation, and k_{nwo} is the permeability to the nonwetting fluid at connate (or irreducible) wetting fluid saturation (Fig. 10). The external boundary conditions are

$$x = 0, \quad p_w = p_1; \qquad x = L, \quad p_{nw} = p_2 \tag{86}$$

Integrating Eqs. (81) and (82) yields

$$p_w = Ax + B \qquad \text{and} \quad p_{nw} = A'x + B' \tag{87}$$

Using conditions (84)–(86),

$$p_w = -\frac{p_1 - p_2}{ML' + (1 - M)\xi}x + p_1, \qquad p_{nw} = \frac{p_1 - p_2}{ML + (1 - M)\xi}M(L - x) + p_2 \tag{88}$$

where

$$M = \frac{k_{wo}}{\mu_w} \Big/ \frac{k_{nwo}}{\mu_{nw}} = \frac{k_{wo}}{k_{nwo}} \cdot \frac{\mu_{nw}}{\mu_w}$$

is called the *mobility ratio*. Hence,

$$q_x = \frac{k_{wo}}{\mu_w} \cdot \frac{p_1 - p_2}{ML + (1 - M)\xi} \tag{89}$$

Since $\partial F/\partial t = -\partial \xi/\partial t$ and $\mathbf{V} \cdot \nabla F = V_x \, \partial F/\partial x = V_x$, we obtain, from Eqs. (71) and (89),

$$-\frac{\partial \xi}{\partial t} + \frac{k_{wo}}{\mu_w n(1 - S_{wo} - S_{nwo})} \frac{p_1 - p_2}{ML + (1 - M)\xi} = 0 \tag{90}$$

Replacing $\partial \xi/\partial t$ by $d\xi/dt$ and solving,

$$t = \frac{\mu_w n(1 - S_{wo} - S_{nwo})}{k_{wo}(p_1 - p_2)} \left[ML\xi + \frac{1}{2}(1 - M)\xi^2 \right] \tag{91}$$

for the time required for the front to advance from $x = 0$ to some distance $x = \xi$. From this it can be seen that the front either accelerates or decelerates, depending on whether $M > 1$ or $M < 1$. In the special case of $M = 1$, the front moves at a constant velocity. Figure 17 shows Eq. (91) in a graphical form [*30*].

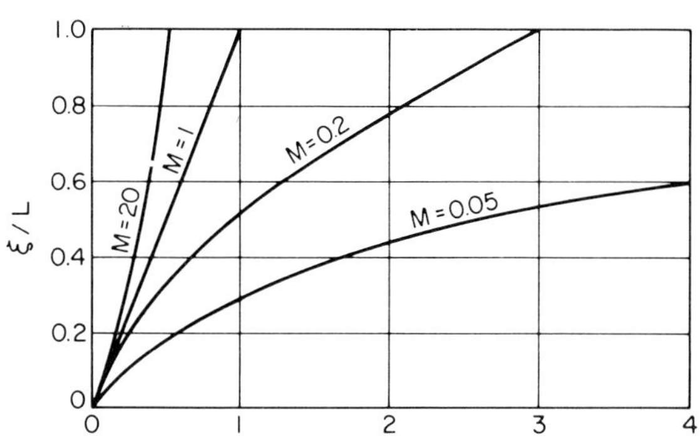

FIG. 17. Linear interface displacement [*30*]. $\tau = \dfrac{k_{wo}\,\Delta p}{\mu_w n(1 - S_{wo} - S_{nwo})L^2}\, t.$

If the medium is inhomogeneous, with $k = k(x)$ and $n = n(x)$, Eqs. (82) and (83) are replaced by

$$\frac{\partial}{\partial x}\left[\frac{k_{wo}(x)}{\mu_w}\frac{\partial p_w}{\partial x}\right] = 0, \qquad \frac{\partial}{\partial x}\left(\frac{k_{nwo}(x)}{\mu_{nw}}\frac{\partial p_{nw}}{\partial x}\right) = 0 \tag{92}$$

and the interface position is determined as before. The boundary conditions may vary with time also. Then, in Eq. (90), $p_1 = p_1(t)$ and $p_2 = p_2(t)$. For complete displacement, insert $k_{wo} = k_{nwo} = k$, and $S_{wo} = S_{nwo} = 0$ in Eqs. (85)–(91). Then $M = \mu_{nw}/\mu_w$, and the time required for the front to reach the boundary $x = L$ is given by

$$t_L = \frac{\mu_w n L^2}{2k(p_1 - p_2)} (M + 1) \tag{93}$$

2. *Radial Encroachment*

Muskat [*30*] and Scheidegger [*1*] present also the solution for radial encroachment where the interface, separating the displacing water (viscosity μ_w) from the displaced oil (viscosity μ_{nw}), has the form of a circle in the horizontal plane advancing from $r = r_e$ to $r = r_w < r_e$. The boundary conditions are $p = p_e$ at $r = r_e$, $p = p_w$ at $r = r_w$. The interface equation is $F \equiv r - \rho(t) = 0$ and the conditions to be satisfied on it are $p_w = p_{nw}$ and $(k_{wo}/\mu_w)\, \partial p_w/\partial r = (k_{nwo}/\mu_{nw})\, \partial p_{nw}/\partial r$. The continuity equation for radial flow of incompressible fluids in a homogeneous isotropic medium is

$$\frac{1}{r}\frac{\partial}{\partial r}\left[\frac{k_{wo}}{\mu_w}\frac{\partial p_w}{\partial r}\right] = 0, \qquad \frac{1}{r}\frac{\partial}{\partial r}\left[\frac{k_{nwo}}{\mu_{nw}}\frac{\partial p_{nw}}{\partial r}\right] = 0 \tag{94}$$

The solution is obtained following the same procedure as that just described for the case of linear displacement. For $k_{wo} = k_{nwo} = k$, $S_w = S_{nw} = 0$, and p_e and p_w independent of time, the solution is

$$\begin{aligned} \frac{4k(p_e - p_w)}{\mu_w r_e^2 n} t = {} & \left(\frac{\rho}{r_e}\right)^2 \left[\ln\left(\frac{\rho}{r_e}\right)^2 - M \ln\left(\frac{\rho}{r_w}\right)^2 + M - 1\right] \\ & - M \ln\left(\frac{r_w}{r_e}\right)^2 - (M - 1) \end{aligned} \tag{95}$$

Other cases given in the literature deal with water drives and water flooding techniques of secondary recovery in oil reservoirs. Scheidegger [*1*] lists some of these works. In most cases, solutions were obtained by numerical or analog methods (see Secs. IV, D and [*24*]).

3. *Interface Displacement in Two-Dimensional Horizontal Flows*

Using the summation convention, Eq. (70), the solution of which gives $F(x, y, z, t)$, can be written in the form

$$\frac{\partial F}{\partial t} - \frac{K^\alpha}{n}\frac{\partial \varphi^\alpha}{\partial x_i}\frac{\partial F}{\partial x_i} = 0 \qquad \left(\text{or,}\quad \frac{\partial F}{\partial t} - \frac{K^\alpha}{n}\nabla\varphi^\alpha \quad \nabla F = 0\right) \tag{96}$$

where $x_1 \equiv x$, $x_2 \equiv y$, $x_3 \equiv z$, and α denotes either the wetting or the non-wetting fluid. Following Muskat [30], let us introduce new orthogonal curvilinear coordinates

$$\xi_1 = \xi_1(x, y, z), \qquad \xi_2 = \xi_2(x, y, z), \qquad \xi_3 = \xi_3(x, y, z)$$

so that

$$\frac{\partial \xi_i}{\partial x_j}\frac{\partial \xi_k}{\partial x_j} = 0$$

Then

$$\frac{\partial \varphi^\alpha}{\partial x_i}\frac{\partial F}{\partial x_i} = \frac{\partial \varphi^\alpha}{\partial \xi_j}\frac{\partial F}{\partial \xi_j}\left[\sum_{(i)}\left(\frac{\partial \xi_j}{\partial x_i}\right)^2\right]$$

If the surface $\xi_1 = \text{const}$ coincides with the equipotential surface $\varphi^\alpha = \text{const}$, then

$$\frac{\partial \varphi^\alpha}{\partial x_i}\frac{\partial F}{\partial x_i} = \frac{\partial F}{\partial \xi_1}\left(\frac{\partial \xi_1}{\partial x_i}\frac{\partial \xi_1}{\partial x_i}\right) \qquad \left(\text{or,} \quad \nabla\varphi^\alpha\ \nabla F = |\nabla \xi_1|^2 \frac{\partial F}{\partial \xi_1}\right) \tag{97}$$

so that Eq. (96) can be written

$$\frac{\partial F}{\partial t} - \frac{K^\alpha}{n}\left(\frac{\partial \xi_1}{\partial x_i}\frac{\partial \xi_1}{\partial x_i}\right)\frac{\partial F}{\partial \xi_1} = 0 \qquad \left(\text{or,} \quad \frac{\partial F}{\partial t} - \frac{K^\alpha}{n}|\nabla \xi_1|^2 \frac{\partial F}{\partial \xi_1} = 0\right) \tag{98}$$

where $|\nabla \xi_1|^2$ has to be expressed as a function of ξ_1, ξ_2, and ξ_3. Equation (98) has the integral

$$F = t + \frac{n}{K}\int \frac{d\xi_1}{|\nabla \xi_1|^2} + g(\xi_1, \xi_2) = \text{const} \tag{99}$$

where g is an arbitrary function to be chosen so that F assumes its initial preassigned form at $t = 0$. For two-dimensional plane flow systems, neglecting the difference in density and viscosity between the two fluids, it is convenient to choose the family of streamlines $\psi = \text{const}$, which is everywhere orthogonal to the curve $\varphi = \text{const}$, as coordinate ξ_1. Essentially we then have the case of steady flow of a homogeneous fluid, but with some fluid surface (say made of marked fluid particles) moving in the flow domain. This possibility of steady flow (i.e., velocity and pressure remain everywhere unchanged with time) yet with a moving interface should be recognized. Then Eq. (99) becomes

$$F(\varphi, \psi, t) = t + \frac{n}{K}\int \frac{d\varphi}{|\nabla \varphi|^2} + g(\psi) = \text{const} \tag{100}$$

in which the integral is to be evaluated along a streamline $\psi = \text{const}$.

Two examples are given by Muskat [30].

a. The flow pattern is that of a single well at a distance d from a line drive lying on the x axis. The x axis is an equipotential ($\varphi = \varphi_0$) boundary and remains so as the interface, originally coinciding with this line, moves towards the pumping well located at $x = 0$, $y = d$. For the steady flow case

$$\varphi_0 - \varphi = \frac{Q}{4\pi K} \ln \frac{x^2 + (y-d)^2}{x^2 + (y+d)^2}$$
$$\psi = -\frac{Q}{2\pi K} \tan^{-1} \frac{-2dx}{x^2 + y^2 - d^2} \tag{101}$$

where Q is the flux per unit thickness into the well. The denominator in the integral of Eq. (100) may then be shown to be

$$|\nabla\varphi|^2 = \frac{Q^2}{4\pi^2 d^2 K^2} (\cosh \xi_1 + \cos \xi_2)^2$$

$$\xi_1 = -\frac{2\pi K(\varphi_0 - \varphi)}{Q}, \qquad \xi_2 = -\frac{2\pi K\psi}{Q}$$

The shape at any time t of the interface composed of fluid particles leaving the line drive at $t = 0$ is given by

$$F(\xi_1, \xi_2, t) = t - n \frac{2\pi d^2}{Q} \int \frac{d\xi_1}{(\cosh \xi_1 + \cos \xi_2)^2} + g(\xi_2) = C = \text{const.}$$

For an initial front which coincides with the x axis $g(\xi_2) = C$. Hence,

$$t = \frac{2\pi d^2 n}{Q \sin^2 \xi_2} \left[\frac{\sinh \xi_1}{\cosh \xi_1 + \cos \xi_2} - 2 \cot \xi_2 \tan^{-1}\left(\tanh \frac{\xi_1}{2} \tan \frac{\xi_2}{2} \right) \right] \tag{102}$$

Figure 18 shows Eq. (102), plotted together with some streamlines. The advancing interface will reach the pumping well at $t = 2\pi n d^2/3Q$ (or, $\tau = 1/3$).

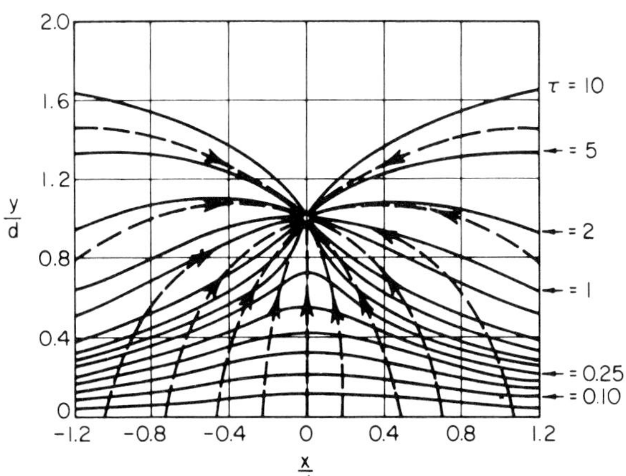

FIG. 18. Advance of a line drive [30]. $\tau = \dfrac{Q}{2\pi n d^2} t$.

The total area swept out by the advancing interface in that time t is $A = Qt/n = 2\pi d^2/3$, or two thirds the area of a circle of radius d.

b. A pumping well and an injection well of equal strength lie along the y axis at $(0, \pm d)$ in an infinite homogeneous flow domain. The interface, composed of fluid particles leaving the injection well at $t = 0$, is advancing from the recharging well toward the pumping well. For this case, the potential distribution and the stream function are also given by Eq. (101). The constant time curves are [30]

$$t = \frac{2\pi n d^2}{Q \sin^2 \xi_2}\left[\frac{\sinh \xi_1}{\cosh \xi_1 + \cos \xi_2} - 2 \cot \xi_2 \tan^{-1}\left(\tanh \frac{\xi_1}{2} \tan \frac{\xi_2}{2}\right) - \frac{\sinh \xi_1{}^0}{\cosh \xi_1{}^0 + \cos \xi_2} + 2 \cot \xi_2 \tan^{-1}\left(\tanh \frac{\xi_1{}^0}{2} \tan \frac{\xi_2}{2}\right)\right] \tag{103}$$

where $\xi_1 = \xi_1{}^0$ is the equipotential defining the injection well at $(0, -d)$. The advancing interface is shown in Fig. 19. In this case the injected fluid will reach the pumping well at $t = 4\pi n d^2/3Q$ $(\tau = 2/3)$.

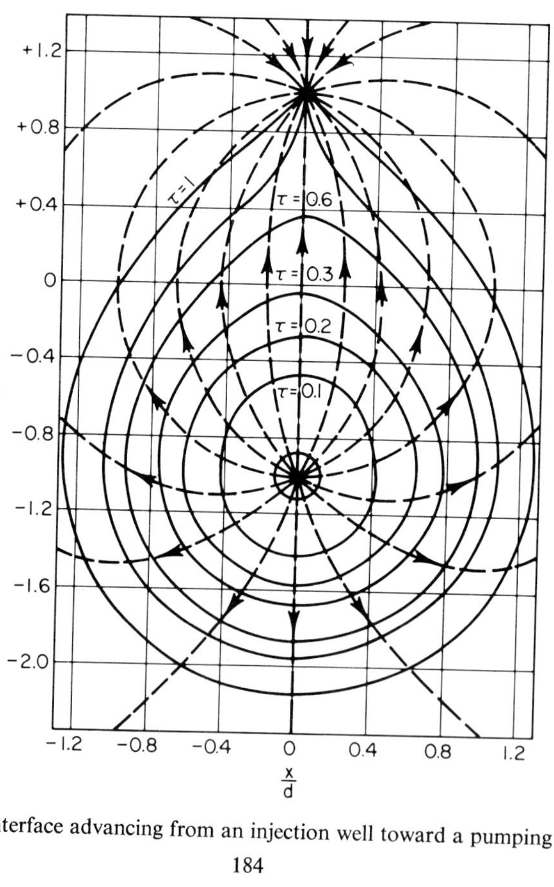

FIG. 19. An interface advancing from an injection well toward a pumping well [30].

A somewhat different approach was suggested by Shchelkachev and Lapuc [4, p. 152]. Using the complex potential,

$$f(z) = \varphi + i\psi; \qquad z = x + iy$$

where

$$\frac{df(z)}{dz} = \frac{\partial\varphi}{\partial x} + i\frac{\partial\psi}{\partial x} = -\frac{1}{K}(q_x - iq_y) = -\frac{n}{K}\frac{d\bar{z}}{dt}$$
$$q_x = n\frac{dx}{dt} = nV_{xj}, \qquad q_y = n\frac{dy}{dt} = nV_y, \qquad \bar{z} = x - iy \tag{104}$$

In the case of incomplete separation n is replaced by $n(1 - S_{wo} - S_{nwo})$. From Eq. (104),

$$t = -\frac{n}{K}\int_{(C)}\frac{d\bar{z}}{df(z)/dz} \tag{105}$$

where the integration is carried out along a streamline. Equation (105) can be derived from Eq. (100) by noting that $|\nabla\varphi|^2 = (df/dz)\overline{(df/dz)}$ and $d\varphi/d\bar{z} = \overline{(df/dz)}$, where the overscore indicates the complex conjugate. The use of this approach is of special interest in the case of complex flow patterns such as those resulting from a large number of sources and sinks. In these cases, the flow pattern $\varphi - \psi$ can be obtained easily by superposition. However, except for a small number of cases the integration has to be carried out numerically. Some examples of the application of Eq. (105) are given in Ref. [4].

Bear and Jacobs [31], by following water particles along streamlines, also determine the shape of an advancing interface separating a body of water injected into a confined aquifer from the indigenous water of the aquifer. Differences in density and viscosity are neglected so that essentially we have here a single homogeneous fluid, a portion of which is labeled as a tracer.

Consider a well at the origin that is injecting tracer-labeled water at a constant rate Q per unit thickness into an infinite homogeneous ($K = \text{const}$) confined aquifer of constant thickness b in which a uniform flow is taking place at a constant specific discharge q_o in the $+x$ direction (Fig. 20). Elastic storativity is neglected and the flow is steady. For this case, the potential distribution and the stream function are given by

$$\begin{aligned}
&\varphi = -(q_o/K)x + (Q/4\pi K)\ln(x^2 + y^2)\\
&\psi = -(q_o/K)y - (Q/2\pi K)\tan^{-1}(y/x)\\
&y = 0, \qquad \theta = \tan^{-1}(y/x) = 0, \qquad \psi = 0\\
&y = 0, \qquad \theta = \pi,\ (x < 0), \qquad \psi = -Q/2K\\
&y = 0, \qquad \theta = 2\pi,\ (x > 0), \qquad \psi = -Q/K
\end{aligned} \tag{106}$$

The velocity components V_x, V_y in the x and y directions, respectively, are

$$V_x = \frac{dx}{dt} = \frac{q_0}{n} + \frac{Qx}{2\pi n(x^2 + y^2)}, \qquad V_y = \frac{dy}{dt} = +\frac{Qy}{2\pi n(x^2 + y^2)} \tag{107}$$

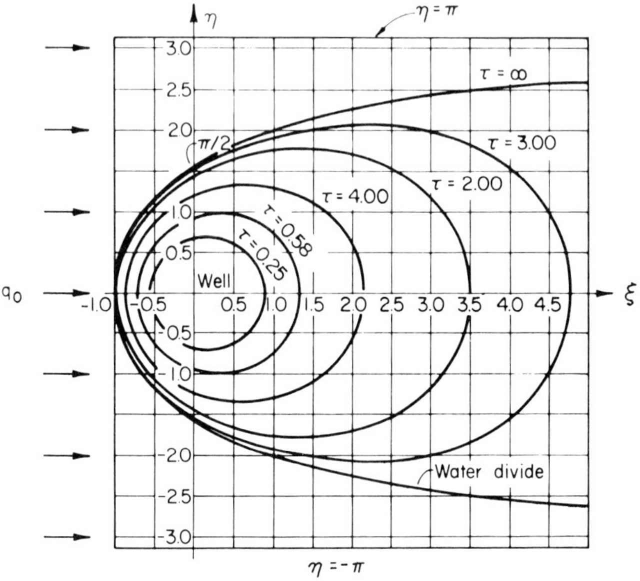

FIG. 20. An advancing interface in the case of a well injecting water into an aquifer with uniform flow [31].

The equations of pathlines (streamlines in steady flow) in the field of flow is

$$dx/V_x = dy/V_y = dt \tag{108}$$

When integrated, these equations yield the parametric forms $x = x(t, t_0)$, $y = y(t, t_0)$ which define the position at time t of a liquid particle that at some previous time t_0 was at a known point. The interface separating the injected water from the indigenous water of the aquifer is composed of all particles leaving the injecting well at the origin at $t_0 = 0$.

By Eq. (106) along a given streamline $\psi = \text{const}$, we have

$$x = y \cot [2\pi(\Psi + q_o y)/Q]; \quad \Psi = K\psi \tag{109}$$

which when inserted into Eq. (107) yields

$$dt = -(2\pi n y/Q) \csc^2 [-2\pi(\psi + q_o y)/Q] \, dy \tag{110}$$

Integrating Eq. (110) for particles leaving $x = 0$, $y = 0$ at $t = 0$, and introducing the dimensionless variables

$$\xi = (2\pi q_o/Q)x, \qquad \eta = (2\pi q_o/Q)y, \qquad \tau = (2\pi q_o{}^2/nQ)t \tag{111}$$

yields

$$\tau = \xi + \ln [\sin \theta/\sin (\eta + \theta)]; \quad \theta = \tan^{-1}(y/x) \tag{112}$$

or

$$\exp (\xi - \tau) = \cos \eta + (\xi/\eta) \sin \eta \tag{113}$$

Equation (112) defines the time required for the front to reach the point (ξ, η). The equation of the front at $\tau = \tau_0$ is

$$\exp\{-\xi[\cos\eta + (\xi/\eta)\sin\eta]\} = \exp(-\tau_0) = \text{const.} \tag{114}$$

The shapes of the advancing front are shown in Fig. 20. As $\tau \to \infty$, we have $\xi = -\eta/\tan\eta$, which is the limiting front position (water divide). The front does not exceed $\eta = \pm\pi$ in the η-direction. For $\tau \to \infty$ and $\eta = 0$, we obtain $\xi = -1$ or $\xi = +1$. The stagnation point $V_x = 0$, $V_y = 0$ is, by Eq. (107), at $\xi = -1$, $\eta = 0$. The front along the ξ axis moves according to

$$\tau = \xi - \ln(1 + \xi) \tag{115}$$

When this method is applied to more complicated flow patterns, numerical methods of integration must be used.

D. Interface Displacement in the Vertical Plane

Consider the case of an inclined homogeneous reservoir of constant thickness (Fig. 21) in which fluid 1 (ρ_1, μ_1) is displacing fluid 2 ($\rho_2 < \rho_1$, μ_2). The displacement is from left to right.

Following Marle [*3*], several cases may be distinguished.

(i). In the absence of flow, the two fluids are separated by a horizontal transition zone as shown in Fig. 21*a*. Below the transition zone, the pore space is filled with fluid 1 and by fluid 2 at residual saturation S_{20}. Above this zone, the pore space is occupied by fluid 2 and by fluid 1 at irreducible saturation S_{10}. Across the transition zone the fluid 1 saturation varies from $(1 - S_{20})$ to S_{10}.

(ii). When flow takes place at a low velocity, experience shows [*32*] that after a certain time a steady flow is established similar to the one-dimensional case, with iso-saturation surfaces in the transition zone taking an inclined position. Two cases are possible: If the mobility ratio $M = (k_{r10}/\mu_1)/(k_{r20}/\mu_2) < 1$, where k_{r10} is the relative permeability to fluid 1 at residual fluid 2 saturation and k_{r20} is the relative permeability to fluid 2 at irreducible fluid 1 saturation, the normal to the iso-saturation surfaces approaches the direction of the flow (Fig. 21*b*). If $M > 1$, the angle between this normal and the direction of the flow becomes larger (Fig. 21*c*).

(iii). If the rate of flow is increased, nothing will change for the case $M < 1$, except that the normal to the iso-saturation surfaces further approaches the direction of the flow, and the displacement is closer to that of one-dimensional flow. If, however, $M > 1$, we observe a new phenomenon. Equilibrium (with a fixed inclination of the iso-saturation surfaces) is possible no longer, and a fluid 1 tongue will penetrate into the region occupied by fluid 2 (Fig. 21*d*). If fluid 1 is water and fluid 2 is oil (water drive) in an oil reservoir, and the movement of the transition zone is toward a producing well, the water

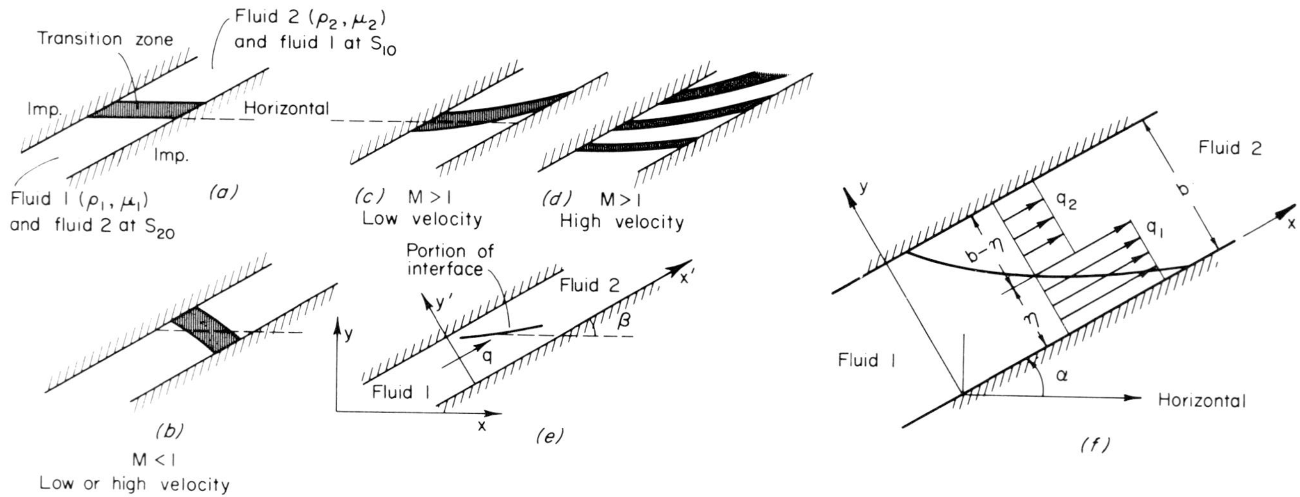

FIG. 21. Effect of mobility and rate of flow on displacement in the vertical plane.

tends to bypass the oil, flowing into the producing well. This flow regime is called *supercritical*; the rate of flow at which the transition occurs is called the *critical rate*.

In all these cases, if the transition zone is relatively narrow, it can be approximated by an abrupt interface. Also, as an approximation, we may sometimes assume that the two fluids are completely separated by the interface ($S_{20} = 0$, $S_{10} = 0$).

In order to determine the critical rate (assuming an abrupt interface), consider the portion of the interface shown in Fig. 21*e*. Below the critical flow rate, the difference in density counterbalances the effect of the difference in viscosity and prevents fluid 1 from bypassing fluid 2. Consequently, both fluids move at the same specific discharge q. Following the discussion leading to Eq. (56) and assuming that $k_1 \neq k_2$ (say because of residual fluid 2 and irreducible fluid 1 saturations),

$$q = -\frac{k_1 \rho_1 g}{\mu_1} \frac{\partial \varphi_1}{\partial x'}, \qquad q = -\frac{k_2 \rho_2 g}{\mu_2} \frac{\partial \varphi_2}{\partial x'}$$

On the interface, equality of pressure leads to

$$\rho_2 \varphi_2 - \rho_1 \varphi_1 = (\rho_2 - \rho_1) y$$

Hence,

$$\rho_2 \frac{\partial \varphi_2}{\partial x'} - \rho_1 \frac{\partial \varphi_1}{\partial x'} = (\rho_2 - \rho_1) \frac{\partial y}{\partial x'}, \qquad q\left(\frac{\mu_2}{k_2} - \frac{\mu_1}{k_1}\right) = g(\rho_1 - \rho_2) \frac{\partial y}{\partial x'} \tag{116}$$

or, since $dy = \sin \beta \cdot dx' + \cos \beta \cdot dy'$,

$$q\left(\frac{\mu_2}{k_2} - \frac{\mu_1}{k_1}\right) = q \frac{\mu_2}{k_2} \left(\frac{M-1}{M}\right) = g(\rho_1 - \rho_2)\left(\tan \beta + \frac{dy'}{dx'}\right) \cos \beta$$

$$M = \frac{(k_1/\mu_1)}{(k_2/\mu_2)} \tag{117}$$

If it is assumed that $q = \text{const}$ (with y'), then $dy'/dx' = \text{const}$ and the interface will assume the shape of a straight line. If $M = 1$, we must have $dy'/dx' = -\tan \beta$, and the interface will be horizontal. If $M > 1$, then $dy'/dx' > -\tan \beta$. If the limiting condition for the possibility of a constant specific discharge in both fluids is given by $dy'/dx' = 0$, then $q = q_{\text{critical}}$ is

$$q_{\text{critical}} = \frac{g(\rho_1 - \rho_2) \sin \beta}{\mu_2/k_2 - \mu_1/k_1} \tag{118}$$

In general, as the interface is displaced, $q_1 \neq q_2$, and Eq. (116) becomes

$$\left(q_2 \frac{\mu_2}{k_1} - q_1 \frac{\mu_1}{k_1}\right) = g(\rho_1 - \rho_2) \frac{\partial y'}{\partial x'} \tag{119}$$

From the discussion in Sec. IV, B, it follows that an exact analytical solution of the shape and position of a moving interface is beyond the scope of this Chapter; several approximate approaches, however, are presented throughout. As an example, a simple approach to the case of an inclined homogeneous isotropic reservoir in which supercritical flow of two incompressible fluids takes place will now be considered.

For an interface that is long with respect to the thickness b of the reservoir (Fig. 21*f*), we may assume in both fluids that flow components in the y direction, i.e., normal to impervious reservoir boundaries, are relatively small and can be neglected. This means that everywhere flow is only in the direction of x. In a way, this is equivalent to the Dupuit assumptions of essentially horizontal flow employed in ground water flows with a phreatic surface. Also assume that within each fluid the specific discharge varies only with x. Obviously, with these assumptions the continuity equation $\operatorname{div} \mathbf{q} = 0$ which is valid for every part cannot be satisfied rigorously. Instead, we shall derive a global continuity equation for each fluid by averaging over the normal to the flow direction.

Accordingly, start from the continuity equation written for a point within the domain occupied by fluid 1, or

$$\frac{\partial q_x}{\partial x} + \frac{\partial q_y}{\partial y} = 0$$

and integrate

$$\int_0^{\eta(x,t)} \frac{\partial q_x}{\partial x}\, dy + \int_0^{\eta(x,t)} \frac{\partial q_y}{\partial y}\, dy = 0$$

Using Leibnitz's rule,

$$\frac{\partial}{\partial x} \int_0^{\eta(x,t)} q_x\, dy - q_x \bigg|_{y=\eta} \frac{\partial \eta}{\partial x} + q_y \bigg|_{y=\eta} - q_y \bigg|_{y=0} = 0 \tag{120}$$

Since

$$q_y \bigg|_{y=0} = 0, \qquad \int_0^{\eta(x,t)} q_x\, dy = Q_{x1}, \qquad q_y \bigg|_{y=n} = n\,\frac{d\eta}{dt}, \qquad q_x \bigg|_{y=\eta} = n\,\frac{dx}{dt}\bigg|_{y=\eta}$$

$$\frac{d\eta}{dt} = \frac{\partial \eta}{\partial t} + \frac{\partial \eta}{\partial x}\frac{dx}{dt}\bigg|_{y=\eta}$$

we obtain, from Eq. (120),

$$\partial Q_{x1}/\partial x + n\, \partial \eta/\partial t = 0 \tag{121}$$

as the continuity equation for fluid 1. In a similar way we obtain for fluid 2,

$$\partial Q_{x2}/\partial x + n\, \partial(b - \eta)/\partial t = 0 \tag{122}$$

In both equations n may be replaced by $n(1 - S_{wo} - S_{nwo})$ in the case of incomplete separation. Eqs. (121) and (122) are called the *shallow depth* (or the *Boussinesq*) *approximate equations.*

Based on the assumptions just stated,

$$Q_{x1} = q_1\eta = -\frac{k_1\rho_1 g}{\mu_1}\frac{\partial\varphi_1}{\partial x}\eta, \qquad Q_{x2} = q_2(b-\eta) = -\frac{k_2\rho_2 g}{\mu_2}\frac{\partial\varphi_2}{\partial x}(b-\eta) \tag{123}$$

where k_1 and k_2 have the same, or different, values depending on whether the separation is complete or incomplete. Also, at the interface, equality of pressure leads to

$$\rho_2\varphi_2 - \rho_1\varphi_1 = (\rho_2-\rho_1)(x\sin\alpha + y\cos\alpha) \qquad \text{at} \quad y=\eta$$

or

$$\rho_2\frac{\partial\varphi_2}{\partial x} - \rho_1\frac{\partial\varphi_1}{\partial x} = (\rho_2-\rho_1)\left(\sin\alpha + \frac{\partial\eta}{\partial x}\cos\alpha\right) \tag{124}$$

Let the total discharge be denoted by Q such that

$$Q = qb = q_1\eta + q_2(b-\eta) = Q_{x1} + Q_{x2} \tag{125}$$

By combining Eqs. (121), (124), and (125),

$$\frac{\partial}{\partial x}\left[\frac{\dfrac{Q\mu_2}{(b-\eta)k^{(2)}} - g(\rho_1-\rho_2)\left(\sin\alpha + \dfrac{\partial\eta}{\partial x}\cos\alpha\right)}{\dfrac{\mu_2}{(b-\eta)k_2} + \dfrac{\mu_1}{\eta k_1}}\right] + n\frac{\partial\eta}{\partial t} = 0 \tag{126}$$

This equation can be rewritten in several forms. For example,

$$\frac{k_1}{\mu_1}\frac{\partial}{\partial x}\left[\eta\,\frac{Q - g(\rho_1-\rho_2)(b-\eta)\left(\sin\alpha + \dfrac{\partial\eta}{\partial x}\cos\alpha\right)\dfrac{k_2}{\mu_2}}{\eta\dfrac{k_1}{\mu_1} + (b-\eta)\dfrac{k_2}{\mu_2}}\right] + n\frac{\partial\eta}{\partial t} = 0 \tag{127}$$

or

$$\frac{\partial}{\partial x}\left[\left(\frac{\eta}{b}\right)\frac{q - g(\rho_1-\rho_2)\left(1-\dfrac{\eta}{b}\right)\left(\sin\alpha + \dfrac{\partial(\eta/b)}{\partial x}\cos\alpha\right)\dfrac{k_2}{\mu_2}}{\dfrac{\eta}{b}\left(1 - \dfrac{k_2}{\mu_2}\Big/\dfrac{k_1}{\mu_1}\right) + \dfrac{k_2}{\mu_2}\Big/\dfrac{k_1}{\mu_1}}\right] + n\frac{\partial\left(\dfrac{\eta}{b}\right)}{\partial t} = 0 \tag{128}$$

For the horizontal reservoir, set $\alpha = 0$ in the foregoing equations. For example, Eq. (128) becomes

$$\frac{\partial}{\partial x}\left[\left(\frac{\eta}{b}\right)\frac{q - g(\rho_1-\rho_2)\left(1-\dfrac{\eta}{b}\right)\dfrac{\partial(\eta/b)}{\partial x}\dfrac{k_2}{\mu_2}}{\dfrac{\eta}{b}\left(1 - \dfrac{k_2}{\mu_2}\Big/\dfrac{k_1}{\mu_1}\right) + \dfrac{k_2}{\mu_2}\Big/\dfrac{k_1}{\mu_1}}\right] + n\frac{\partial\left(\dfrac{\eta}{b}\right)}{\partial t} = 0 \tag{129}$$

Equations (126)–(129) are second-order partial differential equations in the dependent variable η. Their solution yields the shape of the advancing interface: $\eta = \eta(x, t)$. Unfortunately, the equations are nonlinear, and an exact analytical solution even in this approximate form is not possible without further simplifications (see, for example, Sec. VI, D). However, the advantage of using these equations is that the dependent variable η is a function of two independent variables x and t, whereas the original dependent variable φ is a function of three dependent variables x, y, t. The analysis may be extended easily to three dimensions with the elevation of points on the interface expressed by $\zeta = \zeta(x, y, t)$. The resulting equations will be

$$\frac{\partial Q_{x1}}{\partial x} + \frac{\partial Q_{y1}}{\partial y} + n\frac{\partial \zeta}{\partial t} = 0 \tag{121a}$$

$$\frac{\partial Q_{x2}}{\partial x} + \frac{\partial Q_{y2}}{\partial y} + n\frac{\partial (b - \zeta)}{\partial t} = 0 \tag{122a}$$

where $Q_{x1} = \eta q_{x1}$, $Q_{y1} = \eta q_{y1}$, $Q_{x2} = (b - \eta)q_{x2}$, $Q_{y2} = (b - \eta)q_{y2}$.

From Eq. (129) it follows [*3*, p. 9 and *4*, p. 191] that here (η/b) plays the role of a fluid 1 saturation that varies with x, so that the equation is similar to the one describing one-dimensional simultaneous flow of two immiscible fluids (Sec. III, D) in a porous medium where the effective permeability varies linearly with the saturation. This also indicates the possibility of using approximate solution methods mentioned in Sec. III, D. By neglecting $\partial^2\eta/\partial x^2$, we actually assume that the interface has the form of an elongated tongue with a very small curvature. Dietz [*33*] presents an approximate solution based on this assumption. Ref [*4*, pp. 191–212; *32*] also describe several approximate solution methods.

If the two fluids are compressible, the appropriate continuity equations should be used. These may also be averaged over a line normal to the direction of the flow in order to obtain global equations. Also, Eqs. (22) should be used as motion equations.

A different approach to the solution of interface problems that takes into account density and viscosity differences was suggested by Jacquard [*34*], Pilatovski [*35*, *36*], and de Josselin de Jong [*37*]. They sought a single harmonic function in the regions occupied by the two fluids such that it satisfies all external boundary conditions except along the interface itself, the effect of which is taken into account by a singularity property of the function. Pilatovski [*35*, *36*] and Jacquard [*34*] used the *single layer potential* in order to preserve the jump in the normal derivative of the potential perpendicular to the interface. De Josselin de Jong [*37*] replaced the two fluids by a single hypothetical fluid and placed vortices along the interface to account for the change in properties of the actual fluids. The magnitude of vorticity is chosen such that the values of the specific discharge in the hypothetical fluid are everywhere identical to those in the actual fluids. The flow of the hypothetical fluid is determined by the theory of potential flow, using the transformed boundary conditions and the singularities at the vortex points. The magnitude of the vortices for fluids of equal viscosity but different density is proportional to the horizontal component of the density gradient. If the initial position of the interface is known, its subsequent positions can be obtained by first using potential theory to determine the velocity distribution at a certain instant, then displacing the interface according to the velocities at various points along it. This leads to a step-wise method for determining the movement of the interface. Problems with a gradual change in density within a transition zone can be handled in a similar way. Differences in viscosity are accounted for by varying the vorticity according to the discharge which is *a priori* unknown and is part of the required solution. The analysis of this type of problem is much more complicated.

In principle this is an analytic approach, yet the various integrations involved have to be carried out numerically or by means of analogs. Some additional remarks on numerical methods for determining the position of the interface are given in the following section.

E. Numerical and Graphical Methods

When direct analytical solutions of the exact or the approximate equations fail to yield results, numeric or analog methods have to be employed. The latter approach can be implemented in two ways. The first is to solve numerically the equations themselves, once they are established, by such methods as the finite element technique or by replacing the equations that have to be solved by their finite difference approximations. The latter are written in terms of values of the dependent variables at grid points. In general the solution itself is carried out on a high-speed digital computer, and this approach is beyond the scope of this chapter. Nevertheless, it seems appropriate to mention that because of the nonlinearity of the boundary conditions on the interface and owing to the fact that the location of this boundary, even in steady flow with a stationary interface, is *a priori* unknown, it is most difficult to solve interface problems even by numeric techniques and computers.

Numeric techniques are often employed in reservoir engineering to solve problems of simultaneous flow of two fluids in oil reservoirs. Under certain simplifying assumptions, a second numeric approach is possible. This approach is based on the fact that the interface is a material surface, and hence at every instant its motion (in magnitude and direction) is determined by the velocity of fluid particles along it. Once this velocity is known for a sufficient number of particles (points) along the interface, the new position of the front can be obtained by considering the displacements of these fluid particles during a small time interval Δt. The displacement of each particle is given by $\mathbf{\Delta s} = \mathbf{V}\,\Delta t$.

When the mobility ratio is different from unity, the velocity distribution can be obtained from a solution of the various equations and boundary conditions described in the previous sections. As the interface moves, the problem cannot be solved without determining simultaneously the evolution of the interface with time. Thus, even for incompressible fluids the potential distribution is time dependent. One way to solve this difficult unsteady flow problem is to approximate it by a succession of steady states that correspond to successive positions of the interface. At each step a steady state solution is found for that particular interface position. Both the steady state solutions and the interface displacements are often carried out numerically on a digital computer. In many cases analogs [*24*] are used, either to determine the various steady state potential distributions (given the positions of the interface) [*38*], or to determine directly the shape and positions of the moving interface.

The problem is simplified greatly when the mobility ratio [ratio of (k/μ) of displacing fluid behind the front to (k/μ) of the displaced fluid ahead of the front] is equal to unity, or assumed to be approximately so, and gravity effects are neglected (i.e., we consider only the horizontal displacement of the interface as in Sec. IV, C). Under such conditions the flow of the two fluids would occur as if the two fluids were a single homogeneous fluid. The moving interface is then a collection of labeled particles of the same fluid, the velocity of which at every instant may be determined by Darcy's law.

In steady flow it is possible to use potential theory in order to determine the flow pattern, i.e., the equipotentials and the streamlines. In complicated geometries, numerical or analog methods (e.g., the electric analog of the electrolytic tank type) are used to determine the φ-ψ flownet. Once the streamlines in the field of flow have been established, it is possible to determine the movement of the interface by displacing fluid particles belonging to it along streamlines. Thus, if ds denotes an element of length along a streamline, the time required for a fluid particle (belonging to the interface) to travel a length $s(t)$ along the streamline is given by

$$t = \int_0^{s(t)} \frac{ds'}{V_s(s')} = -\frac{n}{K}\int_0^{s(t)} \frac{ds'}{\partial\varphi/\partial s'} \tag{130}$$

where V_s is the velocity in the direction tangent to the streamline. When the potential distribution is obtained numerically or by means of an analog, the computation of t by Eq. (130) is carried out numerically by

$$t = \sum_{(i)} \Delta t = \frac{n}{K} \sum_{(i)} \frac{(\Delta s)^2}{\Delta \varphi} \tag{131}$$

where Δs is an increment of length (along the streamline) along which the potential drop is $\Delta\varphi$.

The incremental displacement during Δt can also be found from

$$\Delta s = \frac{K}{n} \frac{\Delta \psi}{\Delta m} \Delta t \tag{132}$$

where Δm is the distance between two adjacent streamlines. Here, and elsewhere in this chapter, $\Delta\psi = \psi_2 - \psi_1$ between two streamlines is equal to $-\Delta Q/K$.

Of special interest is the case of a number of injection and pumping wells operating simultaneously in a given aquifer or oil reservoir. In this case (still assuming that the mobility ratio is unity), it is necessary to determine the history of advance of the front of the injected fluid. It is also required to determine the instant of *breakthrough*, i.e., the time when the injected fluid reaches the pumping well, and the ratio between the two fluids in the pumped fluid after that instant. Except for simple well arrangements and steady flow (where φ and ψ do not vary with time) [*30*], an analytical solution is most difficult.

As an illustration of a simple numerical solution, consider the following example. Let $A_{i,k}$, $i = 1, 2 \cdots N$, denote N points (fluid particles) each with coordinates $(x_{i,k}, y_{i,k})$ along a moving interface (in the xy horizontal plane) at time $t = k\Delta t$ from the onset of injection. Let points B_j, $j = 1, 2, \cdots M$, with coordinates (x_j, y_j), denote the location of wells in the field. The discharge of each well is Q_j^k (positive or negative) during Δt from $k\Delta t$ to $(k+1)\,\Delta t$. During Δt, all particles $A_{i,k}$ will move to their new positions $A_{i,k+1}$, thus defining the interface at time $t = (k+1)\,\Delta t$. Denoting the distance from $A_{i,k}$ to B_j by $r_{i,k,j}$, the velocity at $A_{i,k}$ caused by well B_j is given by

$$V_{i,k,j} = \frac{\alpha Q_j^k}{r_{i,k,j}}, \qquad \alpha = 1/2\pi n D, \qquad (r_{i,k,j})^2 = (x_{i,k} - x_j)^2 + (y_{i,k} - y_j)^2 \tag{133}$$

where n and D are the porosity and the constant thickness of the confined formation. Summing components in the x and y directions of the elementary displacements of the fluid particles as caused by the simultaneous operation

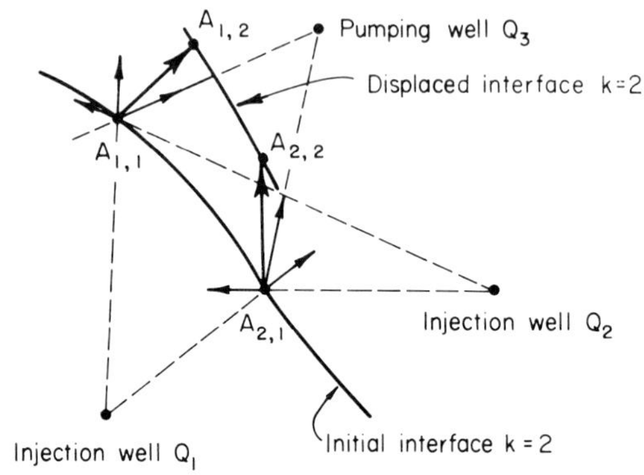

FIG. 22. Graphical method of determining successive interface positions.

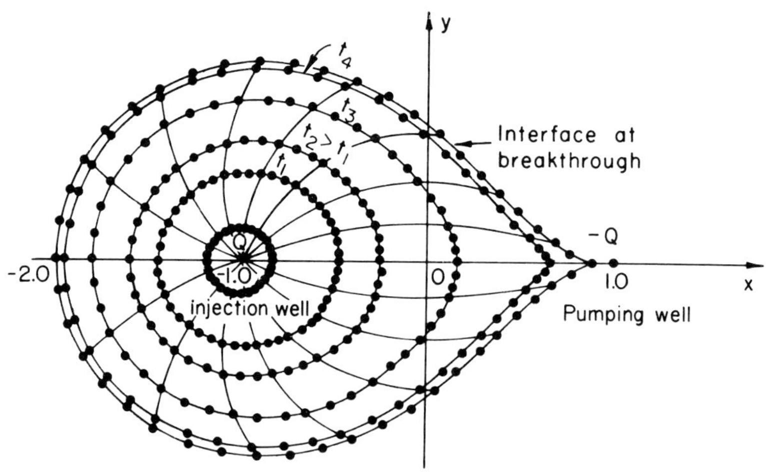

FIG. 23. Interface positions for an injection and a pumping well of equal strength.

of all the wells, we obtain

$$x_{i,k+1} - x_{i,k} = \Delta t \cdot \alpha \sum_{j=1}^{M} \frac{Q_j^k}{(r_{i,k,j})^2} (x_{i,k} - x_j)$$
$$y_{i,k+1} - y_{i,k} = \Delta t \cdot \alpha \sum_{j=1}^{M} \frac{Q_j^k}{(r_{i,k,j})^2} (y_{i,k} - y_j) \qquad (134)$$

In writing Eq. (134), variations of **V** during Δt have been neglected. By repeating this procedure (a digital computer can be used) for a sufficient number of points and for successive time intervals, the history of the interface movement is obtained. In such computations it is always convenient to assume that at time $t = \Delta t$, the front in the vicinity of each injection well of rate Q has the shape of a circle of radius $r_0 = (\alpha Q \, \Delta t)^{1/2}$.

If Q_j^k varies with time, the particles $A_{i,k}$ are displaced approximately along pathlines rather than along streamlines. As Δt is made smaller, a better approximation is obtained. The main difference between this approach and the one described previously is that before particles were displaced exactly along known streamlines, whereas here the moving particles trace streamlines (or pathlines, if Q_j^k varies with time) with an accumulating error. As Δt is reduced, the error becomes smaller. However, the advantage of this approach is that it is not necessary to solve for the $\varphi - \psi$ distribution in the entire field at the end of each time interval.

The procedure described by Eq. (134) can be carried out graphically also by summing elementary displacement vectors. Figures 22 and 23 show the graphical method.

F. Approximate Solutions Based on Linearization

As shown in Sec. IV, B, the interface equations (78) are nonlinear partial differential equations. In order to enable an analytical solution, linearized approximations of these equations have been proposed by various authors. Some of these are described below.

1. *The Small Perturbations Approximation* [*39*, Rept. 3; *40*]

Figure 24 shows an abrupt interface completely separating two immiscible, incompressible liquids in a homogeneous confined aquifer. As the porous medium is homogeneous and isotropic, use is made of the specific discharge potential $\Phi = K\varphi$.

With the nomenclature of Fig. 24, assume that the specific discharge potential Φ in each subregion may be expressed as a sum of a power series of a

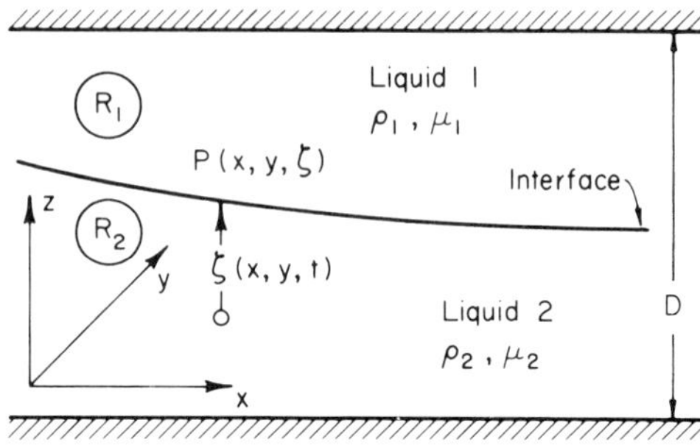

FIG. 24. Interface between two liquids.

small parameter ε:

$$\Phi_i(x, y, z, t) = \Phi_i^\circ(x, y, z) + \varepsilon\Phi_i'(x, y, z, t) + \varepsilon^2\Phi_i''(x, y, z, t) + \ldots, \quad i = 1, 2 \tag{135}$$

where Φ_1° and Φ_2° are steady state specific discharge potential distributions describing an average flow pattern or an initial steady one. The remaining terms of the series in Eq. (135) describe (in each region) deviations from the steady, or average, Φ-distribution, with ε having the character of a *perturbation*. An assumption, commonly used in the solution of similar problems in physics and mathematics, is that $\varepsilon \ll 1$. This assumption is obviously valid when the deviations of the unsteady flow from the steady one are indeed small.

Although the method to be described below (often used in the theory of waves of small amplitude) is also applicable to second [*40*] and higher order linearizations, it will be described and employed here only to derive a first-order linearized solution. This means that only the first two terms on the right-hand side of Eq. (135) will be used.

By inserting Eq. (135) into Eq. (50), and recalling that by definition,

$$\nabla^2\Phi_1^\circ = 0 \quad \text{in} \quad R_1, \qquad \nabla^2\Phi_2^\circ = 0 \quad \text{in} \quad R_2 \tag{136}$$

we obtain

$$\nabla^2\Phi_1' = 0 \quad \text{in} \quad R_1, \qquad \nabla^2\Phi_2' = 0 \quad \text{in} \quad R_2 \tag{137}$$

In a similar manner assume that the elevation ζ of points along the interface can be expanded in a power series, the first two terms of which are

$$\zeta(x, y, t) = \zeta^\circ(x, y) + \varepsilon\zeta'(x, y, t) \tag{138}$$

In Eq. (138), $\varepsilon\zeta'$ (Fig. 25) is the (small) displacement of the interface caused by the unsteady flow, whereas ζ° describes some average, or initial steady state interface position. As is also indicated by the name of this approach, we deal here with only those cases in which the displacement of the interface from

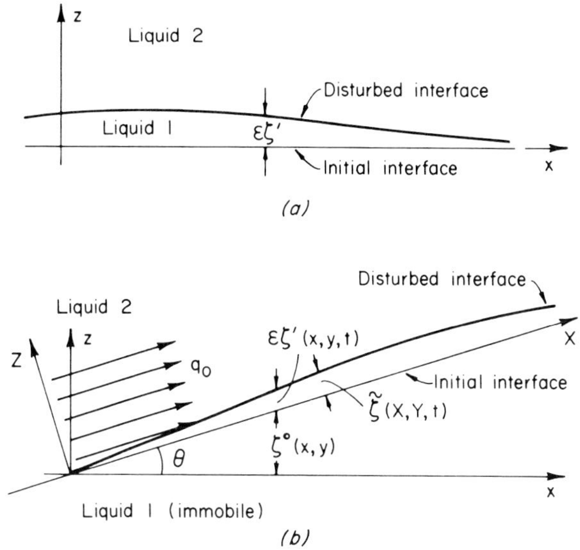

FIG. 25. Initial and disturbed interfaces in (*a*) horizontal and (*b*) inclined flows.

some average or initial position is small with respect to some characteristic length dimension of the flow domain.

Rewriting Eq. (135) for points on the interface ($z = \zeta$) and expanding in a Taylor series,

$$\Phi_i^\circ\bigg|_{z=\zeta=\zeta^\circ+\varepsilon\zeta'} = \Phi_i^\circ\bigg|_{z=\zeta^\circ} + \varepsilon\zeta'\,\frac{\partial\Phi_i^\circ}{\partial z}\bigg|_{z=\zeta^\circ} + O(\varepsilon^2); \quad i = 1, 2$$

Similar expressions may be derived for

$$\Phi_1\bigg|_{z=\zeta}, \qquad \Phi_2\bigg|_{z=\zeta}, \qquad \partial\Phi_1/\partial x\bigg|_{z=\zeta}, \qquad \partial\Phi_1/\partial x\bigg|_{z=\zeta}$$

and for all other expressions appearing in Eq. (78). For example,

$$\partial\Phi_1/\partial x\bigg|_{z=\zeta} = \partial\Phi_1^\circ/\partial x\bigg|_{z=\zeta^\circ} + \varepsilon(\partial\Phi_1'/\partial x + \zeta'\,\partial^2\Phi_1^\circ/\partial x\,\partial z)\bigg|_{z=\zeta^\circ} + O(\varepsilon^2) \quad (139)$$

In this way, all terms are expressed *for points on the average* (or initial) *interface with elevations* $\zeta^\circ(x, y)$.

By introducing these expressions into the boundary conditions of the interface—Eqs. (75) and (78)—and collecting terms of the same power of ε, we obtain the following relationships for ζ', Φ_1' and Φ_2':

$$\zeta' = \frac{(\alpha_2\,\Phi_2' - \alpha_1\Phi_1')}{(1 - \alpha_2\,\partial\Phi_2^\circ/\partial z + \alpha_1\,\partial\Phi_1^\circ/\partial z)} \qquad \text{on} \quad z = \zeta^\circ \qquad (140)$$

$$
\begin{aligned}
&n(\alpha_1\,\partial\Phi_1'/\partial t - \alpha_2\,\partial\Phi_2'/\partial t) - 2\alpha_1[\nabla\Phi_1^\circ\cdot\nabla\Phi_1^\circ + \zeta'\,\nabla\Phi_1^\circ\cdot\nabla(\partial\Phi_1^\circ/\partial z)]\\
&\qquad + \alpha_2[\nabla\Phi_1^\circ\cdot\nabla\Phi_2' + \zeta'\,\nabla\Phi_1^\circ\cdot\nabla(\partial\Phi_2^\circ/\partial z) + \nabla\Phi_2^\circ\cdot\nabla\Phi_1'\\
&\qquad + \zeta'\nabla\Phi_2^\circ\cdot\nabla(\partial\Phi_1^\circ/\partial z)] - (\partial\Phi_1'/\partial z + \zeta'\,\partial^2\Phi_1^\circ/\partial z^2)\\
&\quad = 0 \qquad \text{on} \quad z=\zeta^\circ\\
&n(\alpha_2\,\partial\Phi_2'/\partial t - \alpha_1\,\partial\Phi_1'/\partial t) - 2\alpha_2[\nabla\Phi_2^\circ\cdot\nabla\Phi_2' + \zeta'\,\nabla\Phi_2^\circ\cdot\nabla(\partial\Phi_2^\circ/\partial z)]\\
&\qquad + \alpha_1[\nabla\Phi_2^\circ\cdot\nabla\Phi_2' + \zeta'\,\nabla\Phi_2^\circ\cdot\nabla(\partial\Phi_2^\circ/\partial z) + \nabla\Phi_1^\circ\cdot\nabla\Phi_2'\\
&\qquad + \zeta'\,\nabla\Phi_1^\circ\cdot\nabla(\partial\Phi_2^\circ/\partial z)] + (\partial\Phi_2'/\partial z + \zeta'\,\partial^2\Phi_2^\circ/\partial z^2)\\
&\quad = 0 \qquad \text{on} \quad z=\zeta^\circ
\end{aligned}
\tag{141}
$$

The advantages of replacing Eqs. (75) and (78) by Eqs. (140) and (141) is that the latter are written for the fixed (in time) interface $z = \zeta^\circ$ and that the boundary conditions (141) are *linear* in Φ_1' and Φ_2'.

To find the steady state solution it is necessary to solve Eq. (136) with the boundary conditions

$$
\left.
\begin{aligned}
\zeta^\circ &= \alpha_2\,\Phi_2^\circ - \alpha_1\Phi_1^\circ\\
\alpha_1(\nabla\Phi_1^\circ)^2 - \alpha_2(\nabla\Phi_1^\circ\cdot\nabla\Phi_2^\circ) - \partial\Phi_1^\circ/\partial z &= 0\\
\alpha_2(\nabla\Phi_2^\circ)^2 - \alpha_1(\nabla\Phi_1^\circ\cdot\nabla\Phi_2^\circ) - \partial\Phi_2^\circ/\partial z &= 0
\end{aligned}
\right\} \quad \text{on} \quad z=\zeta^\circ
\tag{142}
$$

Thus the problem has been transformed into one of first solving a steady flow problem for $\Phi_1^\circ(x, y, z)$ and $\Phi_2^\circ(x, y, z)$ satisfying Eq. (142) with a fixed interface of elevation $\zeta^\circ(x, y)$, and then solving Eqs. (137), (140), and (141) for ζ', Φ_1' and Φ_2'. No advantage has been gained with respect to the steady flow problem; the boundary conditions (142) in this case are still nonlinear. They are the same as Eqs. (78), written for steady flow. However, in certain cases the initial steady state solution is known. Again, it should be emphasized that this method of solution is only valid for relatively small interface displacements.

The following two cases are of special interest.

(i). *Initially, uniform flow* (q_o) *takes place in the upper* (γ_1) *liquid, while the lower liquid* (γ_2) *is immobile.* In this case (Fig. 25b), it is convenient to introduce an inclined cartesian coordinate system X, Y, Z, such that the plane $Z = 0$ coincides with the initial planar interface

$$
X = x\cos\theta + z\sin\theta, \qquad Y = y, \qquad Z = -x\sin\theta + z\cos\theta
$$

Then

$$
\Phi_1^\circ = -q_0\,X, \qquad \Phi_2^\circ \equiv 0, \qquad \zeta^\circ = 0, \qquad \tilde{\zeta} = \varepsilon\tilde{\zeta}'
\tag{143}
$$

Also,

$$q_0 = -\partial\Phi_1{}^\circ/\partial X, \qquad \text{i.e.,} \quad \Phi_1{}^\circ = -q_0 X,$$
$$\partial\Phi_2{}^\circ/\partial x \equiv 0, \qquad \text{or} \quad \Phi_2{}^\circ \equiv 0 \tag{144}$$
$$\partial z/\partial x = \sin\theta = \alpha_1 q_0 = \mu_1 q_0/k\Delta\gamma$$

By transforming Eq. (141) from the x, y, z into the X, Y, Z system and expressing $\Phi_1{}^\circ$, $\Phi_2{}^\circ$, $\sin\theta$, and $\cos\theta$ by Eqs. (143) and (144), the following conditions along the interface $Z = \zeta^\circ = 0$ are obtained

$$n(\alpha_1\,\partial\Phi_1{}'/\partial t - \alpha_2\,\partial\Phi_2{}'/\partial t) + (\sin\theta)\,\partial\Phi_1{}'/\partial X$$
$$-\cos\theta\,\partial\Phi_1{}'/\partial Z - (\mu_2/\mu_1)(\sin\theta)\,\partial\Phi_2{}'/\partial X = 0 \tag{145}$$
$$n(\alpha_1\,\partial\Phi_1{}'/\partial t - \alpha_2\,\partial\Phi_2{}'/\partial t) - \cos\theta\,\partial\Phi_2{}'/\partial Z = 0$$

$$\zeta' = (\alpha_2\Phi_2{}' - \alpha_1\Phi_1{}')/\cos\theta \tag{146}$$

Equations (137) in X, Y, Z, together with Eqs. (145) and (146) and appropriate boundary conditions on the external boundaries should yield results for $\Phi_1{}'$ and $\Phi_2{}'$.

(ii). *Both liquids are initially at rest* (*Fig. 25a*). Here $\theta = 0$, $q_o = 0$, $\Phi_1{}^\circ \equiv 0$, $\Phi_2{}^\circ \equiv 0$, $\zeta^\circ = 0$, $\Phi_1 \equiv \varepsilon\Phi_1{}'$, $\Phi_2 \equiv \varepsilon\Phi_2{}'$. Hence, the conditions along the horizontal interface $z = 0$, are

$$n(\alpha_1\,\partial\Phi_1{}'/\partial t - \alpha_2\,\partial\Phi_2{}'/\partial t) - \partial\Phi_1{}'/\partial z = 0; \qquad z = 0$$
$$n(\alpha_1\,\partial\Phi_1{}'/\partial t - \alpha_2\,\partial\Phi_2{}'/\partial t) - \partial\Phi_2{}'/\partial z = 0; \tag{147}$$
$$\zeta' = \alpha_2\Phi_2{}' - \alpha_1\Phi_1{}'$$

Equations (147) have also been derived by Kidder [*42*] and Cailleau, *et al.* [*43*].

It may be interesting to visualize the physical interpretation of Eq. (147). Let the horizontal interface be divided into a large number of squares, each of area A (in the $z = 0$ plane). A closed tube of cross-sectional area $a = nA$ connects the two points: N'' immediately below, and N' immediately above this interface. It is assumed that the initial interface is impervious and that flow through it is possible only through this connecting tube that resists the flow through it. The interface in the connecting tube indicates the position of the moving interface (Fig. 26). An example of the use of this method is given in Sec. VI, E.

Dagan [*44*] showed that for shallow flow ($D/L \ll 1$), the Dupuit approximation used to derive Boussinesq equations (121*a*) and (122*a*) are the first terms in a series expansion, where the small parameter is $(D/L)^2$, or

$$\Phi_i(x, y, z, t) = \Phi_i{}^\circ(x, y, z, t) + (D/L)^2\Phi_i{}'(x\;\; y, z, t); \quad i = 1, 2$$
$$\zeta(x, y, t) = \zeta^\circ(x, y, t) + (D/L)^2\zeta'(x, y, t)$$

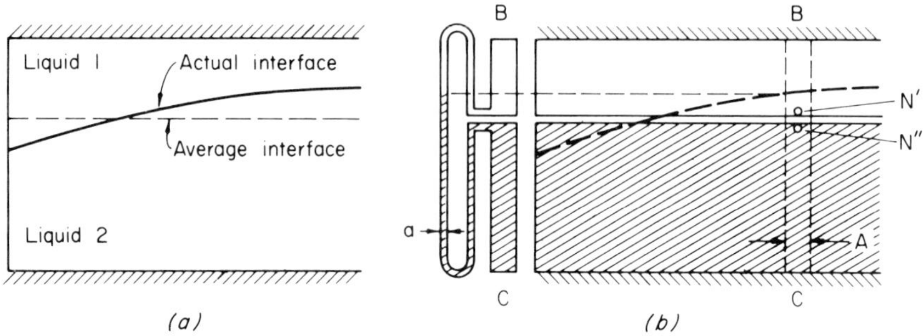

FIG. 26. Physical interpretation of Eq. (147). (*a*) actual conditions, (*b*) the small perturbation approximation.

He then shows that Φ_i° are functions of x, y, and t only.

2. Combining the shallow flow and the small perturbations approximations [39(3)]

This is a linearization of Eqs. (121*a*) and (122*a*) which are based on the Dupuit assumptions. It is assumed that the horizontal aquifer is relatively thin ($D \ll L$) and that the deviations of the interface from its initial position are small with respect to D. Starting from Eqs. (121*a*) and (122*a*), with $q_{x1} = -\partial\Phi_1/\partial x$, $q_{y1} = -\partial\Phi_1/\partial y$, etc., and assuming that ζ, $q_{xi}q_{yi}$, and Φ_i can be expressed as a sum of a steady state term and a deviation,

$$\begin{aligned} \zeta(x, y, t) &= \zeta^\circ(x, y) + \varepsilon\zeta'(x, y, t) \\ q_{xi}(x, y, t) &= \mathring{q}_{xi}(x, y) + \varepsilon q'_{xi}(x, y, t), \quad i = 1, 2 \\ q_{yi}(x, y, t) &= \mathring{q}_{yi}(x, y) + \varepsilon q'_{yi}(x, y, t) \\ \Phi_i(x, y, t) &= \Phi_i^\circ(x, y) + \varepsilon\Phi_i'(x, y, t) \end{aligned} \tag{148}$$

By inserting Eqs. (148) into Eqs. (121*a*) and (122*a*), and neglecting terms of $O(\varepsilon^2)$, the following equations are obtained for the steady state part

$$\begin{aligned} \frac{\partial}{\partial x}[(D - \zeta^\circ)\mathring{q}_{xi}] + \frac{\partial}{\partial y}[(D - \zeta_0)\mathring{q}_{yi}] &= 0 \\ \frac{\partial}{\partial x}[\zeta^\circ \mathring{q}_{x2}] + \frac{\partial}{\partial y}[\zeta^\circ \mathring{q}_{y2}] &= 0 \end{aligned} \tag{149}$$

For the unsteady deviations, the equations are

$$\begin{aligned} n\frac{\partial\zeta'}{\partial t} &= \frac{\partial}{\partial x}[(D - \zeta^\circ)q'_{x1}] + \frac{\partial}{\partial y}[(D - \zeta_0)q'_{y1}] - \frac{\partial}{\partial x}(\zeta'\mathring{q}_{x1}) - \frac{\partial}{\partial y}(\zeta'\mathring{q}_{y1}) \\ -n\frac{\partial\zeta'}{\partial t} &= \frac{\partial}{\partial x}[\zeta^\circ q'_{x2}] + \frac{\partial}{\partial y}[\zeta^\circ q'_{y2}] - \frac{\partial}{\partial x}[\zeta'\mathring{q}_{x2}] - \frac{\partial}{\partial y}[\zeta'\mathring{q}_{y2}] \end{aligned} \tag{150}$$

For the simple case in which both liquids are initially at rest, $q^\circ_{xi} = q^\circ_{yi} \equiv 0$. Then, $\zeta^\circ = \text{const}$, and Eq. (150) becomes

$$n\frac{\partial \zeta'}{\partial t} = (D - \zeta^\circ)\left[\frac{\partial q'_{x1}}{\partial x} + \frac{\partial q'_{y1}}{\partial y}\right], \qquad -n\frac{\partial \zeta'}{\partial t} = \zeta^\circ\left[\frac{\partial q'_{x2}}{\partial x} + \frac{\partial q'_{y2}}{\partial y}\right] \tag{151}$$

Since $q'_{xi} = -\partial\Phi_i'/\partial x$, $q'_{yi} = -\partial\Phi'_{yi}/\partial y$ and $\zeta' = \alpha_2\Phi_2' - \alpha_1\Phi_1'$, Eq. (151) can be written in terms of ζ' only as

$$n\left(\frac{\alpha_1}{D - \zeta^\circ} + \frac{\alpha_2}{\zeta^\circ}\right)\frac{\partial \zeta'}{\partial t} = \frac{\partial^2 \zeta'}{\partial x^2} + \frac{\partial^2 \zeta'}{\partial y^2} \tag{152}$$

Equation (152) is a *linear* partial differential equation of the parabolic type [compare it, for example, with Eq. (129)]. It is analogous to the approximate equation describing flow with a phreatic surface, or

$$\frac{n_e}{KH}\frac{\partial h}{\partial t} = \frac{\partial^2 h}{\partial x^2} + \frac{\partial^2 h}{\partial y^2} \tag{153}$$

where n_e is effective porosity and $h(x, y)$ is the thickness of the saturated region having an average value of H. The resistivity $1/KH$ in Eq. (153) is replaced by $[\alpha_1/(D - \zeta^\circ) + \alpha_2/\zeta^\circ]$ in Eq. (152).

G. Interface Stability

This subject has already been discussed briefly by Scheidegger [*45*], and he introduces mainly the statistical theory of the fingering process (Refs. 15 to 18 in Ref. [*45*]). A few additional comments seem appropriate now to complete the picture of two fluid flows.

Up to this point it was assumed that the abrupt interface (which, as you recall, is a macroscopic approximate concept) was a regular surface separating the two immiscible fluids. However, under certain conditions this approximation fails, and instead of the displacement of the interface (front) as a whole in a regular form, protuberances occur that may advance through the porous medium at velocities much higher than that of the average front. This *instability* (or *fingering*) phenomenon (Fig. 27) always occurs when

$$M = (k_1/\mu_1)/(k_2/\mu_2) > 1 \tag{154}$$

where M is the mobility ratio, k is the permeability, and subscripts 1 and 2 refer to displacing and displaced fluids, respectively. This phenomenon of instability that is initiated by the local *microscopic inhomogeneity* of the medium is most important in oil reservoirs where it reduces sweep efficiencies. There, when oil is displaced by water, k_1 is the permeability to water at residual oil saturation, and k_2 is the permeability to oil at connate water saturation.

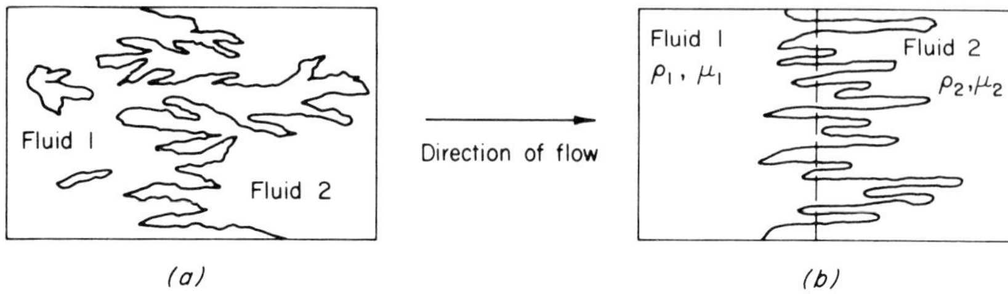

FIG. 27. Instability at a moving interface. (*a*) actual, (*b*) schematic.

Because the frontal instability occurs at a miscroscopic level, its analysis in an exact manner is not possible. The random nature of the microscopic porous structure suggests an approach that involves a statistical description of the porous medium. Chouke *et al.* [*46*, *48*] studied the growth of sinusoidal shaped fingers by studying the wavelength of maximum growth. Scheidegger [*45*] introduces the effect of heterogeneities contained in the porous medium by introducing a random perturbation term into the differential equation of motion for each spectral component of the fingers. By doing so he arrives at an equation describing the growth in time of each spectral component of the fingers. He shows under given external conditions that fingering should be independent of the velocity of the front's displacement. A different approach is to study simplified displacement models that are assumed to represent the fingering phenomenon.

Scheidegger [*1*, p. 202] uses the model of an abrupt front in horizontal flow with maximum saturation of the respective fluids on both sides of the front in order to analyze the phenomenon of fingering. Let Fig. 28*a* represent the

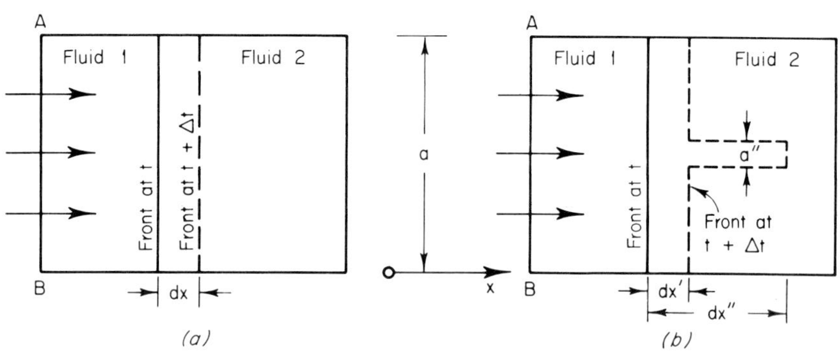

FIG. 28. Analysis of fingering [*1*]. (*a*) stable displacement, (*b*) model of a finger.

horizontal displacement during dt assuming a sharp front and a stable displacement obeying Darcy's law. If V is the average fluid velocity, then

$$dx = V\,dt, \qquad -V = (k_1/\mu_1 n)\,\mathrm{grad}\,p_1 = (k_2/\mu_2 n)\,\mathrm{grad}\,p_2 \tag{155}$$

where n is porosity and p_1 and p_2 are the pressures in the two regions. The displaced volume U is

$$U = na\,dx \tag{156}$$

Suppose that a tiny region on the front is (locally) more permeable than the surrounding region. Then, as the front advances, it will move more rapidly through this more permeable region, giving rise to a small "bump," or "finger." Fig. 28*b* shows how such a "finger" develops. In this case, the displacement caused by the same volume U entering through AB is

$$U = n(a'\,dx' + a''\,dx''); \qquad a'' + a' = a \tag{157}$$

Such a finger can be maintained stable if it has no tendency either to spread out or to retreat and disappear. This is possible only if the pressure gradient in it is the same as in the surrounding liquid. Because $M \neq 1$, we must have $V_1 \neq V_2$. Equation (155) then becomes

$$n(\mu_1/k_1)V_1 = n(\mu_2/k_2)V_2$$

With $V_1 = dx'/dt$, $V_2 = dx''/dt$, and $M = (k_1/\mu_1)/(k_2/\mu_2)$,

$$dx'/dx'' = M \tag{158}$$

According to Scheidegger, a finger will develop only if by its formation it consumes less energy than the corresponding stable displacement of the whole front. In the first case, the work W^A required to move fluid 1 into the space of width dx and fluid 2 out of this space is

$$W^A = a\,|\mathrm{grad}\ p_1|\,(dx)^2 - a\,|\mathrm{grad}\ p_2|\,(dx)^2 = na\,(dx)^2\left(\frac{\mu_1}{k_1} + \frac{\mu_2}{k_2}\right)\left(\frac{dx}{dt}\right) \tag{159}$$

In the second case the work for displacement is

$$W^B = n\left[\frac{\mu_1}{k_1}\,a'(dx')^2 + 2\frac{\mu_2}{k_2}\,a''\,(dx'')^2 + \frac{\mu_2}{k_2}\,a'\,(dx')^2\right]\left(\frac{dx}{dt}\right) \tag{160}$$

From $W^B \leqslant W^A$ Scheidegger shows that fingering (i.e., $dx'' > dx'$) occurs whenever $M > 1$.

A somewhat different approach is presented by Marle [3]. In order to introduce also the effect of density differences, he studies a vertical displacement of a horizontal front (Fig. 29). In the stable case (without a finger),

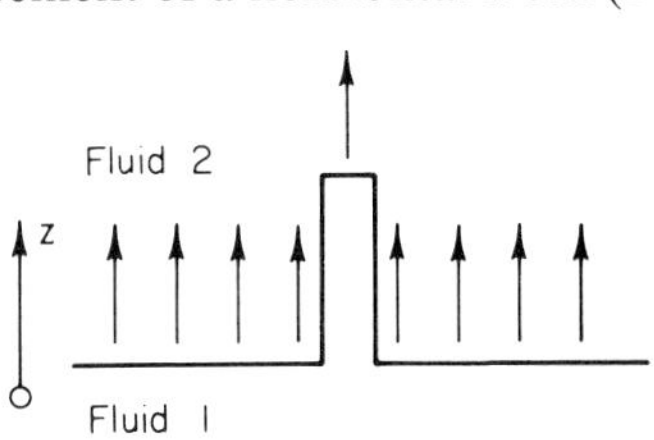

FIG. 29. A finger in the vertical direction [3].

the specific discharges q_1 of fluid 1 and q_2 of fluid 2 are given by

$$q_1 = -\frac{k_1}{\mu_1}\left(\frac{\partial p_1}{\partial z} + \rho_1 g\right), \qquad q_2 = -\frac{k_2}{\mu_2}\left(\frac{\partial p_2}{\partial z} + \rho_2 g\right) \tag{161}$$

The front moves at a velocity $V_F = q_1/n = q_2/n$, where n stands for either the porosity (for complete separation) or for $n\,(1 - S_{wo} - S_{nwo})$ in the case of incomplete separation; k_1 and k_2 are effective permeabilities. When a finger is formed, let V_D, q_D be the velocity and the specific discharge of its edge. Within the finger,

$$q_D = -\frac{k_1}{\mu_1}\left(\frac{\partial p_2}{\partial z} + \rho_1 g\right) = V_D n \tag{162}$$

In Eq. (162), $\partial p_2/\partial z$ is used because the finger is completely surrounded by fluid 2 which imposes its pressure on the fluid within the finger; $V_D > V_F$ means that the finger tends to grow, and vice versa. From Eqs. (161) and (162) it follows that

$$V_D - V_F = -\frac{k_1}{n\mu_1}\left(\frac{\partial p_2}{\partial z} - \frac{\partial p_1}{\partial z}\right) \tag{163}$$

or, using Eq. (161),

$$V_D - V_F = -\frac{k_1}{n\mu_1}\left[nV_F\left(\frac{\mu_1}{k_1} - \frac{\mu_2}{k_2}\right) + g(\rho_1 - \rho_2)\right] \tag{164}$$

The last equation may also be written in the form

$$\begin{aligned} V_D - V_F &= -(1 - M)[V_F + g(\rho_1 - \rho_2)/(n\mu_1/k_1)(1 - M)] \\ &= (M - 1)[V_F - V_C]; \quad V_C = -\frac{k_1 g(\rho_1 - \rho_2)}{\mu_1(1 - M)n} \end{aligned} \tag{165}$$

Let $V_F > 0$ denote vertically upward flow; also $g > 0$ (downward).

(i). Then, if $\mu_2/k_2 < \mu_1/k_1$ (i.e., $M < 1$) and $\rho_1 > \rho_2$, $V_C < 0$, and always $V_D - V_F < 0$, the front is stable.

(ii). If $\mu_2/k_2 < \mu_1/k_1$ (i.e., $M < 0$) and $\rho_1 < \rho_2$, always $V_C > 0$, but there are two possibilities: if $V_F > V_C$, the front is stable; if $V_F < V_C$, the front is unstable.

(iii). If $\mu_2/k_2 > \mu_1/k_1$ (i.e., $M > 1$) and $\rho_1 < \rho_2$, the displacement is stable if $V_F < V_C$ and unstable if $V_F > V_C$.

(iv). If $\mu_2/k_2 > \mu_1/k_1$ (i.e., $M > 1$) and $\rho_1 < \rho_2$, we cannot have $V_F < V_C$ because $V_C > 0$ and the front is always unstable.

The velocity V_C is a certain critical velocity that determines the stability of the front. Jacquard and Seguier [*47*] also investigated this problem reaching similar criteria.

Another force that may affect the stability of a front is the capillary force. Only approximate theories (e.g., Chouke *et al.* [*46*, *48*]) exist on this subject. The capillary forces act mainly in the direction perpendicular to the fingers and tend to equalize saturations. Therefore, these forces always tend to stabilize the displacement.

V. Determining the Steady Interface by the Hodograph Method

A. General

The hodograph method is a powerful mathematical tool for solving steady, two-dimensional problems of flow through porous media, especially where a phreatic surface or an interface between two immiscible liquids comprises a portion of the domain's boundary. As was already mentioned (Sec. IV, B), the main difficulty in solving a flow problem involving an interface as a boundary stems from the fact that the latter's position and shape are *a priori* unknown. In fact, they are the required solution, or at least part of it. The advantage of solving the problem in the hodograph plane is that in this plane the shape of the interface boundary is completely known. In the following paragraphs the hodograph method and some related mappings are discussed mainly in connection with the interface.

Although the literary meaning of the term hodograph is a *graph of velocities*, we shall use it here to mean graph of specific discharge, which is the average (over the entire section) velocity of the flow through the porous medium.

The elements of the hodograph theory were presented by Helmholtz and Kirchhoff in the nineteenth century when they studied discontinuous motion in classical hydrodynamics. Prandtl and Tietjens [5], who attribute the term hodograph to Hamilton, as well as many texts in hydrodynamics, describe the method in connection with two-dimensional free surface flows. Hamel [*49*] developed the method in connection with free surfaces and seepage faces in flow through porous media. Dachler [*50*] introduced the method to ground water flows, Muskat [*30*], who describes the method, also mentions Davison [*51*] as one who applied the method to the problem of an earth dam with vertical walls. Vedernikov [*52*], Polubarinova-Kochina [*53*], Aravin and Numerov [*54*], Harr [*55*], and Bear and Dagan [*56*] describe and apply the hodograph method to various ground water flow problems. Henry [*57*], Bear and Dagan [*39*, Rept. 1; *58*], de Josselin de Jong [*59*], among others, applied the method to the solution of salt-water fresh-water interface problems. They made the simplifying assumption that the two types of water are immiscible liquids and that the interface is an abrupt one. This assumption also underlies the discussion whenever an interface between fresh water and salt water is mentioned in the following paragraphs. Kidder [*60*], Van Quy [*61*], and others applied it to problems encountered in (oil) reservoir engineering.

B. Boundary Conditions

The following boundary conditions occur in problems of steady flow through porous media. The free surface and the seepage face boundaries are briefly reviewed also, as they often occur in a problem that also involves the interface. The discussion is related only to steady two-dimensional flow in the vertical xy plane.

The boundary curve C may be described mathematically by

$$F(x, y) = 0 \tag{166}$$

Let **ln** denote a unit vector normal to the boundary (i.e., to a line tangent to it) at some point P on it, and directed inward. We then have

$$|\nabla F|\,\mathbf{ln} = \nabla F = \frac{\partial F}{\partial x}\,\mathbf{lx} + \frac{\partial F}{\partial y}\,\mathbf{ly}, \qquad |\nabla F|^2 = \left(\frac{\partial F}{\partial x}\right)^2 + \left(\frac{\partial F}{\partial y}\right)^2 \tag{167}$$

where **lx** and **ly** are unit vectors in the x and y directions, respectively. Denoting the angles that **ln** makes with the $+x$ and $+y$ axes by α_{nx} and α_{ny}, respectively,

$$\cos\alpha_{nx} = \frac{(\partial F/\partial x)}{|\nabla F|}, \qquad \cos\alpha_{ny} = \frac{(\partial F/\partial y)}{|\nabla F|} \tag{168}$$

At every point P on C we define a unit vector **ls** in the direction of the tangent to C at P, such that **ls** is obtained from **ln** by a clockwise rotation by 90°.

In all cases we shall assume that the medium is homogeneous and isotropic.

1. *Boundary of Prescribed Potential*

This can be stated as

$$\varphi = \varphi(x, y) \qquad \text{or} \quad \varphi = \varphi(s) \qquad \text{on} \quad C \tag{169}$$

where C denotes the considered portion of the boundary and s is length measured along C.

A special case of Eq. (169) is the *equipotential boundary* where $\varphi = \text{const.}$ In this case, the vector $\mathbf{J} = -\text{grad}\,\varphi$ is perpendicular to the boundary and, therefore, parallel to its normal **ln**. We then have

$$\mathbf{J} \times \mathbf{ln} = 0, \qquad \nabla\varphi \times \mathbf{ln} = 0 \tag{170}$$

where the symbol **x** denotes a cross- (or vector-) product. With Eq. (167), Eq. (170) becomes

$$\frac{\partial \varphi}{\partial x} \cos \alpha_{ny} - \frac{\partial \varphi}{\partial y} \cos \alpha_{nx} = 0 \tag{171}$$

with $\mathbf{q} = K\mathbf{J}$, Eq. (171) may be written

$$-\frac{q_x}{K} \cos \alpha_{ny} + \frac{q_y}{K} \cos \alpha_{nx} = 0 \tag{172}$$

Since $\cos \alpha_{nx} = \cos (90 + \alpha_{sx}) = -\sin \alpha_{sx}$; $\cos \alpha_{ny} = \cos \alpha_{sx}$, we obtain

$$q_y = q_x \tan (\alpha_{sx} - 90^\circ) \tag{173}$$

2. *Boundary of Prescribed Flux*

Along a boundary of this type, the flux q_n, i.e., discharge (per unit length) normal to the boundary, is given as a function of position. With $q_n = (\mathbf{q} \cdot \mathbf{ln})$, this condition is expressed as

$$q_n = q_n(x, y) \qquad \text{on} \quad C \tag{174}$$

or

$$\nabla \varphi \cdot \mathbf{ln} = f(x, y) \qquad \text{on} \quad C \tag{175}$$

We may also express this condition by the stream function Ψ

$$\Psi = \Psi(s) \qquad \text{on} \quad C \tag{176}$$

The stream function (dim L^2/T) Ψ is such that given two streamlines on which $\Psi = \Psi_1 =$ const and $\Psi = \Psi_2 =$ const, the discharge ΔQ between them is given by $\Delta\Psi = \Psi_2 - \Psi_1$. In addition, in an isotropic medium the velocity potential $\Phi = K\varphi$ and the stream function Ψ are related to each other by the Cauchy–Riemann conditions

$$\partial\Phi/\partial x = \partial\Psi/\partial y, \qquad \partial\Phi/\partial y = -\partial\Psi/\partial x \tag{177}$$

In the steady flow of an incompressible liquid considered here, the stream function Ψ also satisfies the Laplace equation ($\nabla^2\Psi = 0$).

A special case of this type of boundary is the *impervious boundary* where the flux vanishes everywhere. Hence,

$$\mathbf{q} \cdot \mathbf{ln} = 0 \quad \text{on} \quad C, \qquad \nabla\varphi \cdot \mathbf{ln} = \partial\varphi/\partial n = 0 \quad \text{on} \quad C$$

$$\frac{\partial \varphi}{\partial x} \cos \alpha_{nx} + \frac{\partial \varphi}{\partial y} \cos \alpha_{ny} = 0 \quad \text{on} \quad C, \qquad -\frac{\partial \varphi}{\partial x} \sin \alpha_{sx} + \frac{\partial \varphi}{\partial y} \cos \alpha_{sy} = 0 \tag{178}$$

or

$$q_y = q_x \tan \alpha_{sx} \tag{179}$$

3. *A Steady Phreatic Surface without Accretion*

We employ here the *approximate concept of a phreatic surface* on which the pressure is atmospheric (say, $p = 0$) and which serves as a boundary of a saturated flow domain. We neglect the presence of any moisture in the void space above the phreatic surface.

With this approximation in mind, we have at all points of the phreatic surface in the vertical xy plane $p = 0$; hence,

$$\varphi = y, \qquad \varphi - y = 0 \tag{180}$$

Since $\varphi = \varphi(x, y)$, and by Eq. (180) the expression

$$F(x, y) \equiv \varphi - y = 0 \tag{181}$$

is valid for all points on the steady phreatic surface without accretion, $F(x, y) = 0$ may serve as a specification of the boundary's (= phreatic surface) geometry. Actually $\varphi(x, y)$, and hence also the shape of the phreatic surface, is not known until the problem is solved. Being a steady interface between air and water, it behaves like an impervious boundary for which Eq. (178) holds.

From Eqs. (178) and (181) we obtain the boundary condition for two-dimensional flow in the xy plane in the form

$$\nabla\varphi \cdot \nabla F = \nabla\varphi \cdot (\nabla\varphi - \mathbf{ly}) = 0$$

or

$$\left(\frac{\partial\varphi}{\partial x}\right)^2 + \left(\frac{\partial\varphi}{\partial y}\right)^2 - \left(\frac{\partial\varphi}{\partial y}\right) = 0, \qquad (\nabla\varphi)^2 - \partial\varphi/\partial y = 0 \tag{182}$$

It is interesting to note that this condition (like the Laplace equation) is independent of K. The equation describing the steady phreatic surface in an isotropic medium is purely a geometric affair; it depends only on the geometry of the flow domain (see and compare Sec. IV, B).

4. *The Steady Phreatic Surface with Accretion*

Accretion (N) is the rate (volume per unit horizontal area per unit time) at which liquid is added to, or withdrawn from, a saturated region of porous medium at its air-liquid interfaces. We shall refer to it as positive when liquid is added to the saturated region: $\mathbf{N} = -N\mathbf{ly}$. For a stationary interface, the requirement of continuity leads to

$$N \cdot \mathbf{ln} = \mathbf{q}_n \qquad \text{or} \qquad (\mathbf{N} - \mathbf{q}) \cdot \mathbf{ln} = 0$$

from which we obtain

$$(\mathbf{N} - \mathbf{q}) \cdot \mathbf{1n} = (\mathbf{N} - \mathbf{q})(\nabla \varphi - \mathbf{1y}) = 0$$

or, for an isotropic medium,

$$(-N\mathbf{1y} + K\,\nabla\varphi)(\nabla\varphi - \mathbf{1y}) = 0$$

$$\left(\frac{\partial \varphi}{\partial x}\right)^2 + \left(\frac{\partial \varphi}{\partial y}\right)^2 - \frac{\partial \varphi}{\partial y}\frac{K + N}{K} + \frac{N}{K} = 0 \tag{183}$$

5. *The Seepage Surface*

A seepage face always occurs (Muskat [*30*], Polubarinova-Kochina [*53*]), along the external boundary of the flow domain between the downstream end of a phreatic surface and the upper end of a downstream boundary of open water. Along the seepage face, water emerges from the porous medium into the external space and trickles downward on it. Being part of the external boundary, its geometry is, in general, known, except for its upper end which also belongs to the (*a priori*) unknown phreatic surface. Neglecting a thin layer of water flowing above it, the seepage face is exposed to atmospheric pressure. Hence, Eq. (180) remains valid along the seepage face. Since the seepage face is an equipressure surface, we have along it

$$\nabla p \times \mathbf{1n} = 0 \tag{184}$$

where

$$\nabla p = \gamma(\nabla \varphi - \mathbf{1y}) \qquad \text{and} \qquad \mathbf{1n} = \nabla F / |\nabla F|$$

and $F(x, y) = 0$ is the equation of the seepage face boundary.

Equation (180) may therefore be written

$$(\nabla \varphi - \mathbf{1y}) \times (\cos \alpha_{nx}\, \mathbf{1x} + \cos \alpha_{ny}\, \mathbf{1y}) = 0$$

$$\frac{\partial \varphi}{\partial x} \cos \alpha_{ny} - \left(\frac{\partial \varphi}{\partial y} - 1\right) \cos \alpha_{nx} = 0$$

$$\frac{\partial \varphi}{\partial x} \cos \alpha_{sx} + \frac{\partial \varphi}{\partial y} \sin \alpha_{sx} = \sin \alpha_{sx}$$

For an isotropic medium, this becomes

$$q_y = q_x \tan(\alpha_{sx} - 90^\circ) - K \tag{185}$$

6. *The Steady Interface between a Moving Liquid γ_1 and a Stationary one ($\gamma_2 > \gamma_1$)*

Setting $\Phi_2 = \text{const}$, $\partial\Phi_1/\partial t = 0$, and $\partial\Phi_2/\partial t = 0$ in Eq. (78), with y replacing z, we obtain the condition

$$\alpha_1\left[\left(\frac{\partial\Phi_1}{\partial x}\right)^2 + \left(\frac{\partial\Phi_1}{\partial y}\right)^2\right] + \frac{\partial\Phi_1}{\partial y} = 0 \tag{186}$$

For an isotropic medium, this becomes

$$\alpha_1({q_x}^2 + {q_y}^2) - q_y = 0$$

or

$${q_x}^2 + {q_y}^2 - q_y K_1' = 0, \qquad {q_x}^2 + (q_y - K_1'/2)^2 = (K_1'/2)^2 \tag{187}$$

When the upper liquid (γ_1) is stationary, while the lower one (γ_2) is flowing, we have, from Eq. (78),

$$\alpha_2\left[\left(\frac{\partial\Phi_2}{\partial x}\right)^2 + \left(\frac{\partial\Phi_2}{\partial y}\right)^2\right] - \frac{\partial\Phi_2}{\partial y} = 0$$

or

$${q_x}^2 + {q_y}^2 + q_y K_2' = 0 \tag{188}$$

7. *Outflow Face of Liquid* 1 *into a Reservoir of Stationary Liquid* 2 ($\gamma_2 > \gamma_1$)

The boundary condition along a surface of this type is $\varphi_2 = \text{const}$. However, since all boundary conditions are referred to φ_1 (of the flowing liquid 1), we have to use Eq. (75) in order to express the boundary condition in terms of φ_1 by eliminating p. We obtain

$$\varphi_1 = \varphi_2 \frac{\gamma_2}{\gamma_1} - y\,\frac{\gamma_2 - \gamma_1}{\gamma_1} \tag{189}$$

where $\varphi_2 = \text{const}$. For example, for $\varphi_2 = 0$,

$$\varphi_1 = -y(\gamma_2 - \gamma_1)/\gamma_1$$

which is not an equipotential boundary in terms of φ_1. We may also write the boundary condition

$$\nabla\varphi_2 \times \mathbf{1n} = 0 \tag{190}$$

which, in view of Eq. (189), may be written

$$\left(\frac{\gamma_1}{\gamma_2}\nabla\varphi_1 + \mathbf{1y}\frac{\Delta\gamma}{\gamma_2}\right) \times \mathbf{1n} = 0$$

$$\left(\nabla\varphi_1 + \mathbf{1y}\frac{\Delta\gamma}{\gamma_1}\right) \times \mathbf{1n} = 0$$

$$\frac{\partial\varphi_1}{\partial x}\cos\alpha_{ny} - \left(\frac{\partial\varphi_1}{\partial y} + \frac{\Delta\gamma}{\gamma_1}\right)\cos\alpha_{nx} = 0$$

or

$$q_y = q_x \tan(\alpha_{sx} - 90^\circ) + K_1' \tag{191}$$

C. Description of Boundaries in the Hodograph Plane

Let C be some arc in a two-dimensional field of flow (Fig. 30*a*) in the xy plane—the *physical plane*. At each point along this arc there exists a certain specific discharge vector $\mathbf{q}$ with components q_x and q_y in the x and y directions, respectively ($\mathbf{q}_1$ at A, $\mathbf{q}_2$ at B, etc.). Now consider another plane with a Cartesian coordinate system q_x and q_y (parallel to $+x$ and $+y$, respectively (Fig. 30*b*). From 0′ we now draw the vectors $\mathbf{q}_1$, $\mathbf{q}_2$, etc. The end points (A', B', etc.) of these radius vectors form a curve C' that describes the specific discharge (in magnitude and direction) at the corresponding points A, B, etc., along the curve C in the xy plane. We say that the curve C' is the *hodograph* (or *hodograph representation*) of curve C of the given flow. We may also regard this transformation as a *mapping* procedure. Points, curves, and regions in the xy plane are mapped onto corresponding points, curves, and regions in the hodograph plane.

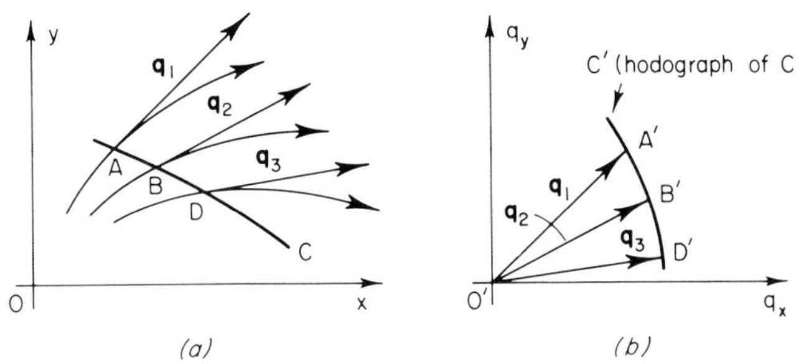

Fig. 30. Hodograph representation of a curve. (*a*) physical plane, (*b*) hodograph plane.

Of special interest is the mapping of the boundaries of a given flow domain in the xy plane. Using the complex variable $z = x + iy$, where $i = \sqrt{-1}$, to denote a point in this plane, we may also refer to the xy plane as the z plane. (See texts on the theory of functions of complex variables for information on this topic.) This mapping is the first step in the hodograph method. Usually, the boundaries of a flow domain are composed of several straight or curved segments, and along each segment one of the conditions listed in Sec. V, B, exists. With the knowledge of these conditions, it is possible to obtain the hodograph of the various boundary segments. Points in the z plane at which adjacent segments meet each other are mapped onto points in the hodograph plane at which the hodographs of these segments intersect each other. In most cases, although it is possible to obtain the hodograph of a boundary segment, it is impossible to identify points along it as being hodograph representations of specific points in the z plane. However, this is not necessary in the hodograph method.

When the entire boundary of a flow domain (including segments at infinity) are mapped onto the hodograph plane, it encloses a certain portion of this plane. If, in the flow domain no specific discharge (in magnitude and direction) occurs more than at one point, each point of the flow domain is represented by a unique point in the hodograph plane, and vice versa, each point in the hodograph plane represents a unique point in the z plane. Whenever the same specific discharge vector occurs at more than a single point in the z plane, we have a double-valued (or sometimes a multivalued) hodograph plane, and the mapping in the hodograph plane takes the form of a Riemann surface [59]. Let us now briefly review the mapping onto the hodograph plane of the various boundaries listed in Sec. V, B.

1. *An Equipotential Boundary*

For a curved boundary of this type, α_{sx} varies from point to point so that the ratio q_y/q_x varies. However, for a boundary that is a straight line making an angle $\alpha_{sx} = \text{const}$ with the $+x$ axis, it follows from Eq. (173) that the hodograph is a *straight line* through the origin, making an angle $(\alpha_{sx} - 90°)$ with the $+q_x$ axis [i.e., normal to the boundary in the z plane (Fig. 31)].

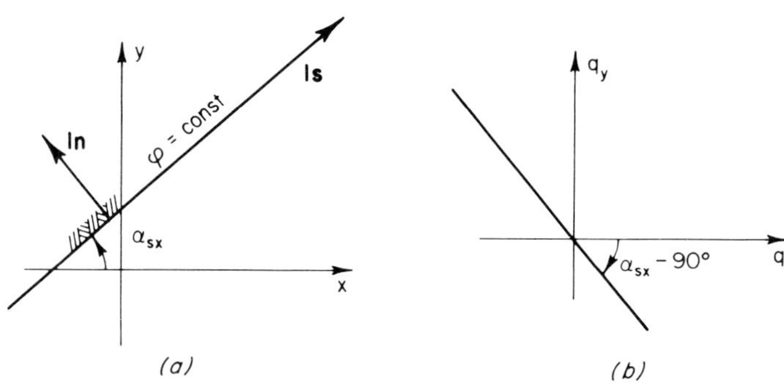

FIG. 31. Hodograph of an equipotential boundary (isotropic medium). (*a*) physical plane, (*b*) hodograph plane.

2. *An Impervious Boundary*

From Eq. (179) it follows that when the impervious boundary has the form of a straight line making an angle α_{sx} with the $+x$ axis, its representation in the hodograph plane is a straight line passing through the origin and making an angle α_{sx} with the $+q_x$ axis, i.e., a line parallel to the boundary in the z plane (Fig. 32).

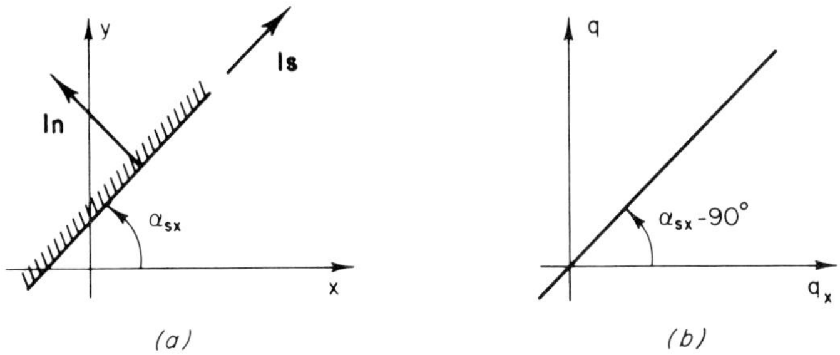

FIG. 32. Hodograph of an impervious boundary. (*a*) physical plane, (*b*) hodograph plane.

3. *A Phreatic Surface without Accretion*

Equation (182) can be rewritten

$$q_x^2 + q_y^2 + q_y K = 0$$

or

$$q_x^2 + (q_y + K/2)^2 = (K/2)^2 \tag{192}$$

In the hodograph plane, this is an equation of a circle of radius $K/2$ centered at $(0, -K/2)$ (Fig. 33).

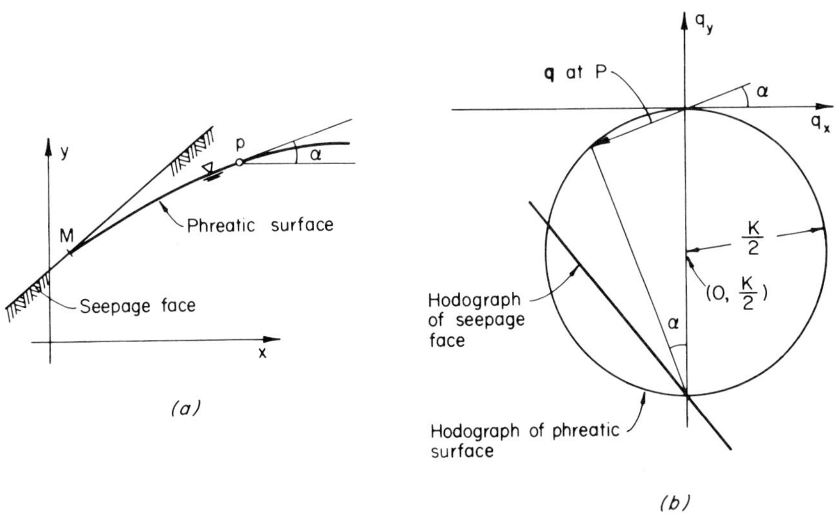

FIG. 33. Hodographs of a phreatic surface and seepage face (isotropic medium). (*a*) physical plane, (*b*) hodograph plane.

4. *A Phreatic Surface with Accretion*

Equation (183) can be rewritten in the form

$$\left(\frac{q_x}{K}\right)^2 + \left(\frac{q_y}{K}\right)^2 + \frac{q_y}{K} \cdot \frac{K+N}{K} + \frac{N}{K} = 0$$

or

$$q_x^{\,2} + \left(q_y + \frac{K+N}{2}\right)^2 = \left(\frac{K-N}{2}\right)^2 \tag{193}$$

This is an equation of a circle of radius $|K - N|/2$ centered at $-(K+N)/2$ (Fig. 34).

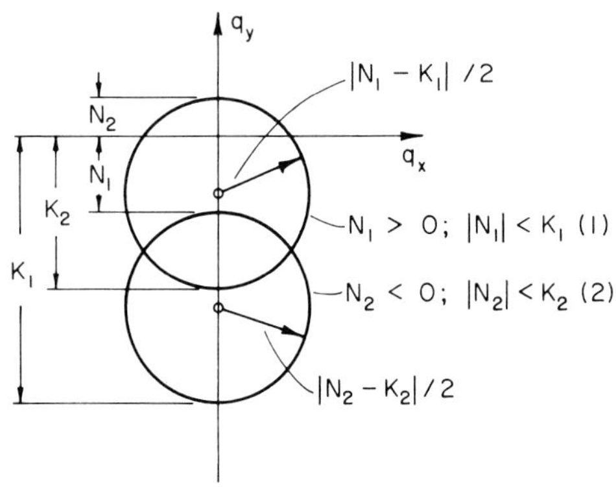

FIG. 34. Hodograph of phreatic surface with accretion (isotropic medium).

5. *A Seepage Surface*

From Eq. (185) it follows that for a seepage surface in the form of a straight line that makes an angle α_{sx} with the $+x$ axis, the hodograph representation is a straight line passing through $(0, -K)$ and making an angle $(\alpha_{sx} - 90°)$—i.e., normal to the seepage surface in the z plane—with the $+q_x$ axis (Fig. 33).

6. *A Steady Interface*

From Eq. (187) it follows that when liquid 1 ($\gamma_1 < \gamma_2$) is moving the hodograph representation of the stationary interface is a circle of radius $K_1'/2$, centered at $(0, K_1'/2)$. For a moving liquid 2, it follows from Eq. (188) rewritten as

$$q_x^{\,2} + (q_y + K_2'/2)^2 = (K_2'/2)^2 \tag{194}$$

that the hodograph representation is a circle of radius K_2' centered at $(0, -K_2'/2)$. This case is analogous to the case of a phreatic surface.

7. *An Outflow Face*

From Eq. (191) it follows that for a straight line outflow face that makes an angle α_{sx} with the $+x$ axis, the hodograph is a straight line passing through $(0, K_1')$ and making an angle $\alpha_{sx} - 90°$ with the $+q_x$ axis.

Three examples of hodograph mappings are shown in Fig. 35. No further explanations seem necessary. It is interesting to note that at point A, the

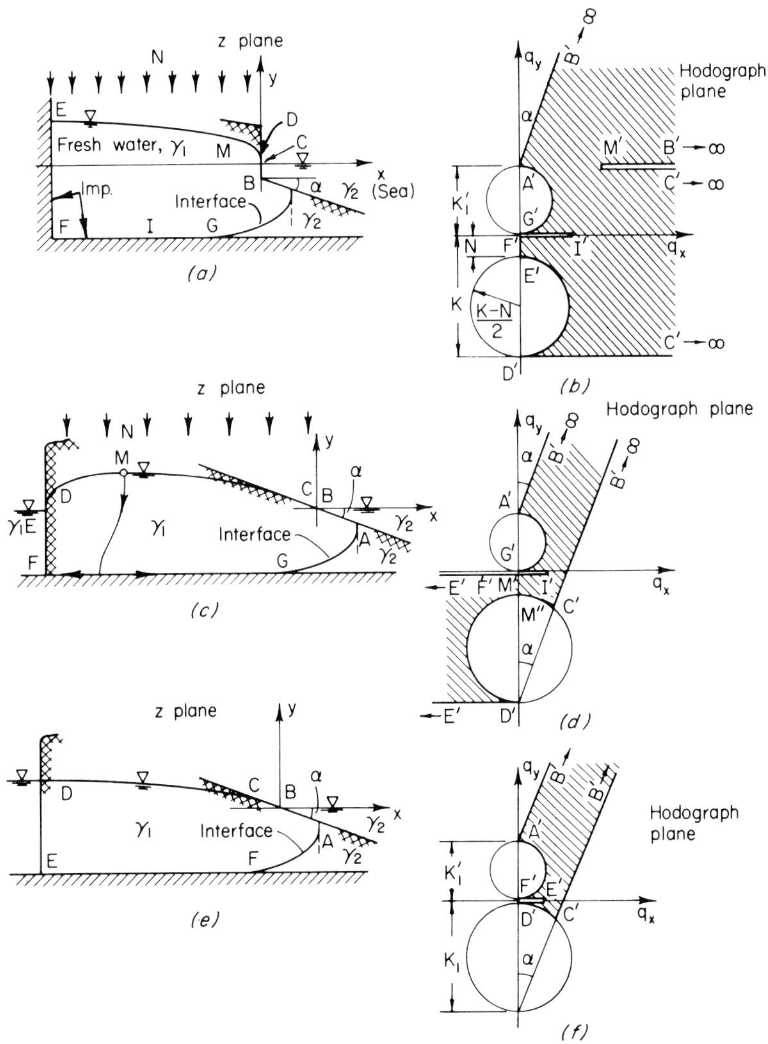

FIG. 35. Examples of hodograph representation of boundaries.

tangent to the interface is *vertical.* The examples correspond to a coastal aquifer under various conditions. Note that in a coastal aquifer $K_1' \ll K_1$.

D. The Hodograph Method

Consider the complex numbers

$$\omega = q_x + iq_y, \qquad \tilde{\omega} = q_x - iq_y \tag{195}$$

where $i = \sqrt{-1}$, ω is the *complex specific discharge*, and $\tilde{\omega}$ is the *conjugate complex specific discharge*. In the same way as the complex number $z = x + iy$ describes a point in the xy plane, the complex number ω describes a point (q_x, q_y) in the hodograph plane (Sec. V, C), and the number $\tilde{\omega}$ describes a point in a plane whose coordinates are $(q_x, -q_y)$. Geometrically, the conjugate $\tilde{\omega}$ of ω is the reflection of ω in the axis of reals (Fig. 36). The $\tilde{\omega}$ plane

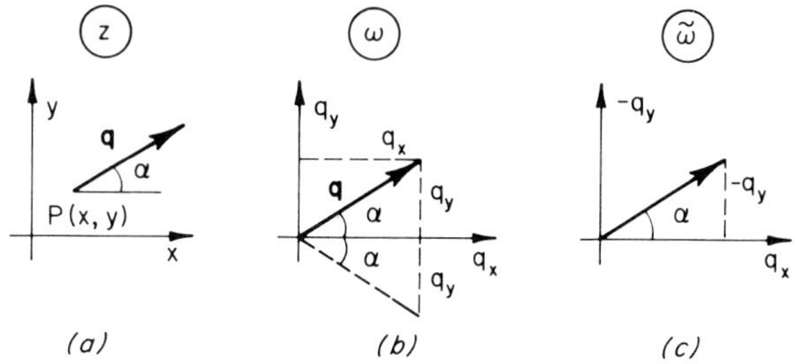

Fig. 36. Definition of z, ω, and $\tilde{\omega}$ planes.

also is often referred to as the hodograph (or sometimes, the inverse hodograph) plane. Here we will refer to it as the hodograph plane, as only this plane is used in the solution of problems by the hodograph method.

The expression $\omega = q_x(x, y) + iq_y(x, y)$ is not an analytic function, whereas $\tilde{\omega}$ is. Expressing $\tilde{\omega}$ in terms of a velocity (or specific discharge) potential $\Phi = K_1\varphi_1$,

$$\tilde{\omega} = \omega_1 + i\omega_2 = q_x - iq_y = -\partial\Phi/\partial x + i\,\partial\Phi/\partial y \tag{196}$$

It can be verified easily that $\tilde{\omega}$ is analytic, as its real ($\omega_1 \equiv q_x$) and imaginary ($\omega_2 = -q_y$) parts, together with their first derivatives, are continuous, single valued (we assume), and satisfy the Cauchy–Riemann conditions

$$\partial\omega_1/\partial x = \partial\omega_2/\partial y, \qquad \partial\omega_1/\partial y = -\partial\omega_2/\partial x \tag{197}$$

We may now relate $\tilde{\omega}$ to the *complex potential* ζ defined by

$$\zeta = f(z) = \Phi(x, y) + i\Psi(x, y)$$

where Φ is the velocity potential and Ψ is the stream function. The function $\zeta = f(z)$ is also analytic. For a single-valued function that is analytic at a point, the derivative at that point is unique, i.e. independent of the direction dz along which the derivative is taken. Hence,

$$\frac{d\zeta}{dz} = \frac{\partial\zeta}{dx}\bigg|_{iy=\text{const}} = \frac{\partial\Phi}{\partial x} + i\frac{\partial\Psi}{\partial x}$$

$$\frac{d\zeta}{dz} = \frac{\partial\zeta}{\partial(iy)}\bigg|_{x=\text{const}} = -i\frac{\partial\Phi}{\partial y} + \frac{\partial\Psi}{\partial y}$$

Applying the Cauchy–Riemann conditions to Φ and Ψ,

$$-\frac{d\zeta}{dz} = q_x - iq_y = \tilde{\omega}, \qquad \left|\frac{d\zeta}{dz}\right| = (q_x{}^2 + q_y{}^2)^{1/2} = |\mathbf{q}| = q \tag{198}$$

From these expressions, it follows that

$$\tilde{\omega}(\zeta) = -d\zeta/dz, \qquad dz = -d\zeta/\tilde{\omega}(\zeta), \qquad z = -\int d\zeta/\tilde{\omega}(\zeta) + \text{const} \tag{199}$$

which relates ζ (and hence Φ and Ψ) to z. We see here that an essential element of this method is the relationship between ζ and $\tilde{\omega}$. With $\Phi = \Phi(x, y)$ and $\Psi = \Psi(x, y)$, we may consider another plane—the ζ plane with coordinates Φ and Ψ—onto which the z plane is mapped. As was just shown, it is always possible to map the z plane onto the ω plane and then, by reflection, onto the $\tilde{\omega}$ plane. Hence, whenever enough information is available to map the z plane onto the ζ plane, the relationship between ζ and $\tilde{\omega}$ can be obtained by mapping one plane onto the other. However, if not enough information is available to obtain the mapping of the flow domain in the ζ plane, the hodograph method cannot be applied.

The following steps summarize the procedure for solving a flow problem by the hodograph method.

(i) Given the contour C of the boundary of a flow domain, use the rules given in Sec. V, C to map it onto the ω plane.

(ii) By reflecting this contour in the real q_x (axis), obtain the contour Γ of C in the hodograph $\tilde{\omega}$ plane. Since the mapping of C onto Γ is conformal, the sense of rotation around C and around Γ is the same.

(iii) Map the boundary C on the ζ plane. The mapped region in the ζ plane will usually have the form of an infinite strip or a rectangle.

(iv) Determine a conformal mapping relationship $\tilde{\omega} = \tilde{\omega}(\zeta)$ that maps the $\tilde{\omega}$ plane on the ζ plane. In most cases this cannot be done directly, and an intermediate plane (or several such planes) is required. For example, we may map the flow domain in the ζ plane onto the upper half of some $\xi = \mu + i\nu$

plane by $\tilde{\omega} = \tilde{\omega}(\xi)$. We then map the flow domain in the ζ plane on the same upper half of the ξ plane by $\zeta = \zeta(\xi)$. Finally, we establish the required relationship $\tilde{\omega} = \tilde{\omega}(\zeta)$.

(v) Using Eq. (199), we integrate

$$z = \int_{\zeta=\zeta_0}^{\zeta} \frac{d\zeta'}{\tilde{\omega}(\zeta')} = x + iy, \qquad \zeta_0 = \Phi_0 + i\Psi_0, \qquad \zeta = \Phi + i\Psi \tag{200}$$

By comparing real and imaginary parts, we obtain the parametric equations

$$x = x(\Phi, \Psi), \qquad y = y(\Phi, \Psi) \tag{201}$$

from which the solution:

$$\Phi = \Phi(x, y), \qquad \Psi = \Psi(x, y) \tag{202}$$

can be obtained. By performing the integration in Eq. (200) along the portion of the boundary corresponding to the interface, we obtain its shape in the z plane in the form of the parametric equations (201).

Because the boundary of the flow region in the $\tilde{\omega}$ plane always involves a portion of a circle, which is the hodograph of the interface, it is most convenient to employ the inversion transformation

$$\tilde{\omega}' = K'/\tilde{\omega} \tag{203}$$

which maps the $\tilde{\omega} = \omega_1 + i\omega_2$ plane onto the $\tilde{\omega}' = \omega_1' + i\omega_2'$ such that

$$\omega_1' = K'\omega_1/(\omega_1^2 + \omega_2^2), \qquad \omega_2' = -K'\omega_2/(\omega_1^2 + \omega_2^2)$$

Hence, straight lines $\omega_2 = a\omega_1$ are transformed into straight lines $\omega_2' = -aK'\omega_1'$; the circle $\omega_2 = a\omega_1 + b (b \neq 0)$ into the circle $(\omega_1' + K'a/2b)^2 + (\omega_2' + K'/2b)^2 = (K'/2b)^2(1 + a^2)$ and the circle $\omega_1^2 + (-\omega_2 + K'/2)^2 = (K'/2)^2$ into the straight line $\omega_2' = 1$. In this way, the hodograph representation of the flow domain in the absence of seepage faces and sloping outflow faces becomes a polygon that can be mapped onto the upper half of some ξ plane by the Schwarz–Christoffel transformation [*62*]. With the inverse transformation of Eq. (203) and mapping both ζ and $\tilde{\omega}$ onto the upper half of an auxiliary plane, Eq. (199) becomes

$$z = -\int \frac{\tilde{\omega}'(\xi)}{K'} \left(\frac{d\zeta}{d\xi}\right) d\xi + \text{const} \tag{204}$$

E. Examples

Five examples are given to demonstrate the application of the hodograph method.

Example 1. An interface in a confined aquifer. Consider [58] the interface in an infinite coastal confined aquifer of constant thickness D (Fig. 37*a*) in which fresh water (γ_1) is discharged to the sea (γ_2) through a horizontal outflow face AB above a stationary interface BE. Figures 37*b* and *c* show the

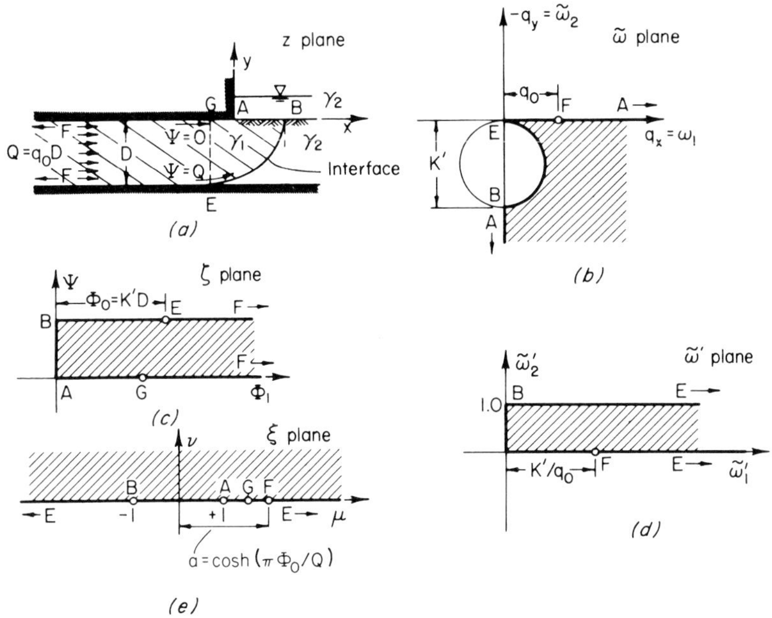

FIG. 37. Determining the interface in a confined aquifer by the hodograph method.

flow domain $FGABEF$ mapped onto the $\tilde{\omega}$ and ζ planes, respectively. Figure 37*d* shows the flow domain in the $\tilde{\omega}'$ plane, where $\tilde{\omega}' = K'/\tilde{\omega}$. Note that at point A we set (arbitrarily) $\Phi = \Phi_1(=K\varphi_1) = 0$. At point $E, y = -D, p = \gamma_2 D$, $\Phi_2 = 0$; $\Phi = \Phi_1 = K'D$.

The flow regions in both the $\tilde{\omega}$ and the $\tilde{\omega}'$ planes are now mapped onto the upper half of the ξ plane by

$$\xi = \cosh \pi\tilde{\omega}' \tag{205}$$

$$\zeta = \frac{Q}{\pi} \cosh^{-1} \frac{a\xi - 1}{a - \xi} \tag{206}$$

where $a = \cosh(\pi K'D/Q) = \pi K'/q_0$. Using Eq. (199) and Eqs. (204)–(206),

$$dz = -\frac{d\zeta/d\xi}{\tilde{\omega}(\xi)}\, d\xi = -\frac{Q(a^2-1)^{1/2}\cosh^{-1}\xi}{\pi^2 K'(a-\xi)(\xi^2-1)^{1/2}}\, d\xi \tag{207}$$

To derive the parametric equations of the interface in the z plane, we integrate Eq. (207) along the segment BE. In the ξ plane this means integration in the range $-\infty < \xi < -1$. Rewriting Eq. (207) with $\eta = \cosh^{-1}(-\xi)$,

$$dz = \frac{Q}{\pi^2 K'} \frac{(a^2 - 1)^{1/2}}{a + \cosh \eta} (\eta + i\pi)\, d\eta \tag{208}$$

Then, by separating real and imaginary parts,

$$dx = \frac{Q}{\pi^2 K'} \frac{(a^2 - 1)^{1/2}}{a + \cosh \eta} \eta\, d\eta \tag{209}$$

$$dy = \frac{Q}{\pi K'} \frac{a^2 - 1}{a + \cosh \eta} d\eta \tag{210}$$

By integrating Eq. (210)

$$y = \frac{Q}{\pi K'} F_1(\eta)$$

where

$$F_1(\eta) = \ln \frac{a + 1 + (a^2 - 1)^{1/2} \tanh(\eta/2)}{a + 1 - (a^2 - 1)^{1/2} \tanh(\eta/2)} \tag{211}$$

For large values of a, y may be approximated by

$$y = \frac{Q}{\pi K'} \eta - \frac{Q}{\pi K'} \ln \frac{\exp(\eta) + 2a}{1 + 2a}$$

or

$$y = \frac{Q}{\pi K'} [\eta - F_3(\eta)], \qquad F_3(\eta) = \ln \frac{\exp(\eta) + 2a}{1 + 2a} \tag{212}$$

From Eq. (209),

$$x = \frac{Q}{\pi^2 K'} F_2(\eta), \qquad F_2(\eta) = F_1(\eta) - \int_0^\eta F_1(\eta')\, d\eta' \tag{213}$$

The values of F_2 may be obtained by numerical integration. Thus, Eqs. (212) and (213) are the parametric equations (parameter η) of the interface. Figure 38 shows various possible shapes of the interface in a confined aquifer as obtained from these equations.

For comparison, Fig. 38 includes two additional curves: one is obtained by using the Dupuit assumptions (Sec. VI, D) with $y = 0$ at $x = 0$; the other is the parabola obtained by Glover [*63*] by using the complex potential $\zeta = \Phi + i\Psi$ satisfying

$$z = \zeta^2 K'/2Q \tag{214}$$

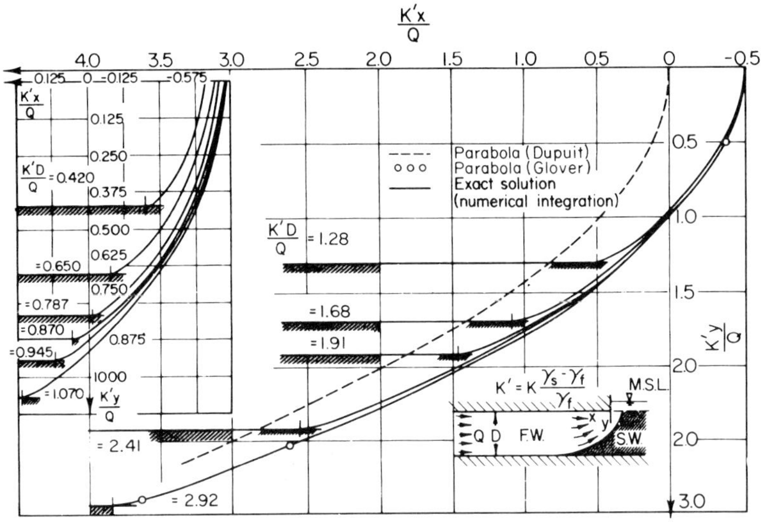

FIG. 38. Shapes of an interface in a confined aquifer [58].

for an aquifer of infinite depth. By separating real from imaginary parts we obtain

$$\begin{aligned}\varphi &= [x + (x^2 + y^2)^{1/2}]^{1/2}(Q/K')^{1/2} \\ \psi &= [-x + (x^2 + y^2)^{1/2}]^{1/2}(Q/K')^{1/2}\end{aligned} \tag{215}$$

where ψ is the conjugate of φ such that $K_1\psi = \Psi$. The interface then takes the form of the parabola

$$y^2 = (2Q/K')x + (Q/K')^2 \tag{216}$$

with $y = 0$, $x = x_0 = -Q/2K'$.

Two additional conclusions may be drawn from the exact solution just derived. The first concerns the length of the outflow face (AB in Fig. 37a)

$$\overline{AB} = \int_{-1}^{+1} \frac{Q(a^2 - 1)^{1/2}}{\pi^2 K'} \frac{\cosh^{-1} \xi}{(a - \xi)(\xi^2 - 1)^{1/2}} \, d\xi$$

which for sufficiently large values of $K'D/Q$ $(K'D/Q > 1.3)$ becomes

$$\overline{AB} = Q/2K'$$

This result agrees with Glover's solution [Eq. (216)] for an aquifer of infinite depth.

The second conclusion concerns the piezometric head $\varphi(G)$ at point G located on the upper confining layer directly above the interface toe E. This point will be of special interest in Sec. VI. From $q_x = -\partial\Phi/\partial x$ and with $q_y = 0$ along AG, we obtain from

$$\tilde{\omega}' = q_x - iq_y = q_x = \pi K'/\cosh^{-1}\xi = -\partial\Phi/\partial x$$

$$\Phi(G) = K\phi(G) - \int_o^{L-x_0} (\pi K'/\cosh^{-1}\xi)\, dx \tag{217}$$

$$= [Q(a^2-1)^{1/2}/\pi] \int_o^{\beta_G} \{1/(a-\cosh\beta)\}\, d\beta$$

where $\beta = \cosh^{-1}\xi$ and β_G is obtained from

$$\overline{AG} = \int_o^{L-x_0} dx = [Q(a^2-1)/\pi^2] \int_o^{\beta_G} \{\beta/(a-\cosh\beta)\}\, d\beta \tag{218}$$

Example 2. Upconing toward a sink (drain) in an aquifer of infinite extent in all directions [58]. Consider a point sink A of constant strength Q located at some distance above an interface (Fig. 39*a*) between fresh water (γ_1) and salt water (γ_2). When upconing takes place, this distance is denoted by b. Figure 39*b* shows the $\tilde{\omega}$ plane of the left half of the flow domain in the z plane.

An inflection point B, where the velocity makes an angle $\beta (0 < \beta < \pi/2)$ with the horizontal axis, exists somewhere along the interface. The case $\beta = \pi/2$ (Fig. 39*d*) corresponds to the limit case for which salt water is still at rest. The case $\beta = \pi/2$ is called *critical situation* in the following discussion. The corresponding discharge is called the *critical discharge* Q_{cr}. The hodograph representation of half the flow region in the critical situation is shown in Fig. 39*e*. Figure 39*c* shows the ζ plane representation for the general case of upconing. The flow region in the $\tilde{\omega}$ plane is now mapped on the upper half of the $\xi = \mu + i\nu$ plane (Fig. 39*f*) by

$$\tilde{\omega} = -iK'\left[1 - \frac{1}{2}\cot\beta\left\{\left(\frac{m(1+\xi)}{\xi}\right)^{1/2} - \left(\frac{\xi}{m(1+\xi)}\right)^{1/2}\right\}\right]^{-1} \tag{219}$$

$$m = \tan^2(\pi/2 + \beta/2)$$

The strip in the ζ plane is mapped on the same upper half of the ξ plane by

$$\zeta = \frac{Q}{2\pi}\ln\xi \tag{220}$$

By Eq. (199),

$$\frac{dz}{d\xi} = \frac{Q}{2\pi iK'}\frac{1}{\xi}\left\{1 - \frac{1}{2}\cot\beta\left[\left(\frac{m(1+\xi)}{\xi}\right)^{1/2} - \left(\frac{\xi}{m(1+\xi)}\right)^{1/2}\right]\right\} \tag{221}$$

$$z = \frac{Q}{2\pi iK'}\left[\ln\frac{\xi}{[\xi^{1/2} + (1+\xi)^{1/2}]^2} + m^{1/2}\left(\frac{1+\xi}{\xi}\right)^{1/2}\cot\beta\right] + C \tag{222}$$

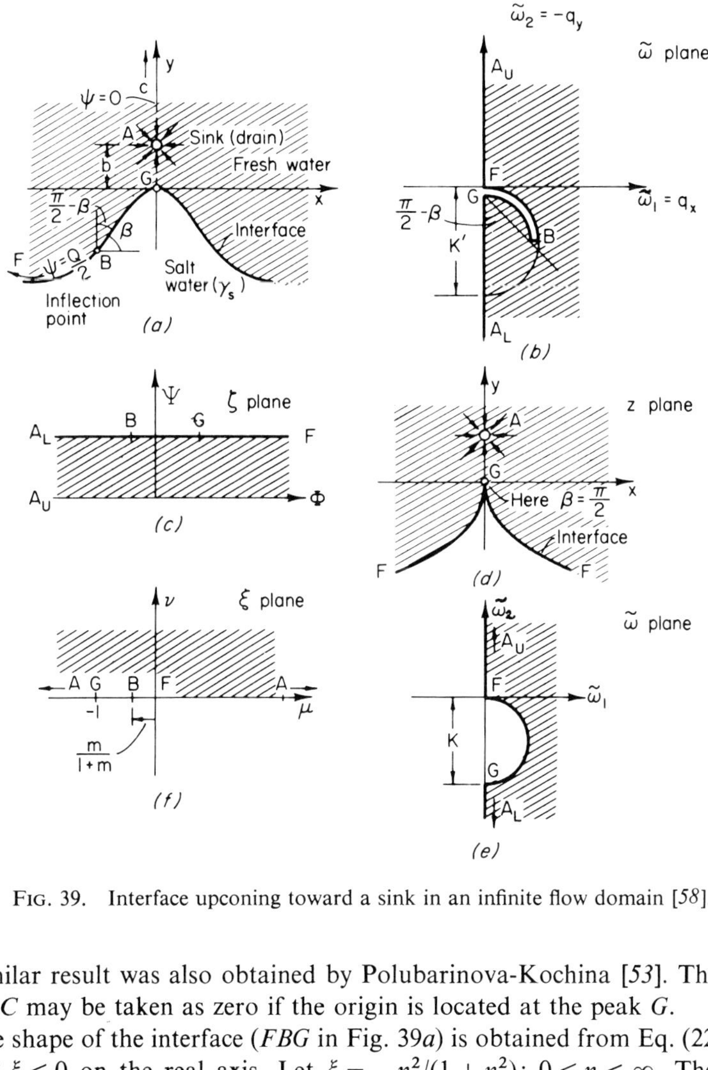

FIG. 39. Interface upconing toward a sink in an infinite flow domain [*58*].

A similar result was also obtained by Polubarinova-Kochina [*53*]. The constant C may be taken as zero if the origin is located at the peak G.

The shape of the interface (FBG in Fig. 39a) is obtained from Eq. (222) for $-1 \le \xi \le 0$ on the real axis. Let $\xi = -\eta^2/(1+\eta^2)$; $0 \le \eta < \infty$. Then the parametric equations of the interface take the form:

$$\frac{xK'\pi}{Q} = -\tan^{-1}\eta - \frac{m^{1/2}\cot\beta}{2}\cdot\frac{1}{\eta} + \frac{\pi}{2} \tag{223}$$

$$\frac{yK'\pi}{Q} = -\frac{1}{2}\ln\frac{\eta^2}{1+\eta^2} \tag{224}$$

The distance b from the top of the upconed interface (Fig. 39*a*) to the sink is obtained by integrating Eq. (221) between $\xi = -\infty$ and $\xi = -1$, or

$$\pi K'b/Q = \tfrac{1}{2}(\ln 4 - m^{1/2} \cot \beta) \tag{225}$$

At the critical situation $\beta = \pi/2$ and

$$\pi K' b_{cr}/Q = 0.097 \tag{226}$$

Figure 40 shows the shapes of the upconed interface for various values

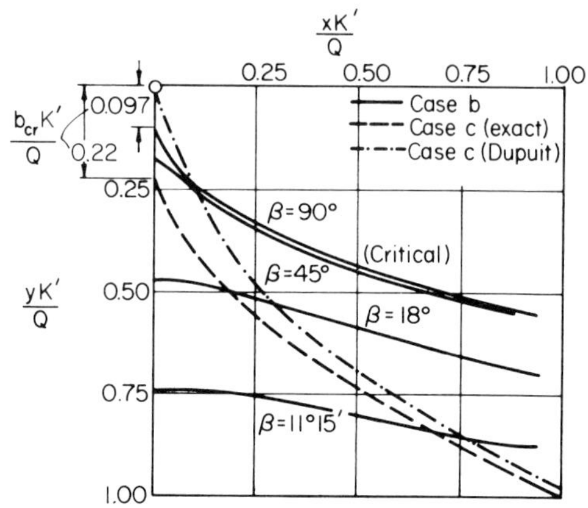

FIG. 40. Shapes of interface below an operating sink [58].

of β. To determine the value of β in a specific case, we must know one point of the interface in the (x, y) plane and solve for β from Eqs. (223) and (224). The value of Q_{cr} may be derived in a similar manner.

For the case of a sink (drain) in an aquifer of infinite thickness, but confined from above by an impervious boundary on which the sink is located, the mapping functions are

$$\begin{aligned} \tilde{\omega}' &= 2a(1 - \xi)^{1/2} + (2/\pi) \tanh^{-1} (1 - \xi)^{1/2} \\ \zeta &= -(Q/2\pi) \ln [\xi/(1 - \xi)] \end{aligned} \tag{227}$$

Example 3. A sink (drain) located on the impervious ceiling of an infinitely thick aquifer, withdrawing fresh water above a fresh water-salt water interface [58]. Figure 41*a* shows the physical z plane with the fresh water discharge Q originating at $x = -\infty$ flowing to the drain A. The interface FBG has an inflection point B where the tangent to the interface makes an angle β with the $+x$ axis. Figures 41*b*, *c*, and *d* show the corresponding $\tilde{\omega}$, $\tilde{\omega}'$, and ζ planes.

The ζ and $\tilde{\omega}'$ planes are mapped onto the upper half of the ξ plane by

$$\zeta = \frac{Q}{\pi} \ln \left(\frac{1}{\xi^*} - 1 \right) \tag{228}$$

and

$$\tilde{\omega}' = -\frac{1}{\pi} \ln \xi^* - \frac{1}{\pi a} (\xi^* - 1) \tag{229}$$

where ξ^* is the conjugate of ξ.

Hence, by using the fundamental relationship between ζ, $\tilde{\omega}$, and z,

$$\frac{dz}{d\xi^*} = -\frac{1}{\zeta} \frac{d\zeta}{d\xi} = -\frac{\tilde{\omega}'}{K'} \frac{d\zeta}{d\xi^*} \tag{230}$$

From Eq. (228),

$$\frac{d\zeta}{d\xi^*} = \frac{Q}{\pi} \frac{1}{\xi^*(\xi^* - 1)} \tag{231}$$

Hence,

$$\frac{dz}{d\xi^*} = \frac{Q}{\pi^2 K'} \cdot \frac{a \ln \xi^* + (\xi^* - 1)}{a\xi^*(\xi^* - 1)} \tag{232}$$

If it is required that $\xi^*(B) = -a$, Eq. (229) yields

$$\cot \beta = -\frac{1}{\pi} \ln a + \frac{1}{\pi a} (a + 1) \tag{233}$$

Because of stability requirements, β cannot exceed $\pi/2$; $\beta > \pi/2$ means that a stationary heavy liquid occupies a region above the lighter liquid. This is an unstable situation. In the limit, for $\beta = \pi/2$, we have $a = 3.59$.

The parametric equations of the interface, which corresponds to $-\infty < \mu < 0$ in the ξ plane, are

$$\frac{dy}{d\mu} = -\frac{Q}{\pi K'} \frac{1}{\mu(\mu - 1)} \tag{234}$$

$$\frac{dx}{d\mu} = \frac{Q[a \ln |\mu| + (\mu - 1)]}{\pi^2 K' a \mu(\mu - 1)} \tag{235}$$

If $\mu = -\eta$ within the range $0 < \mu < +\infty$,

$$\frac{dy}{d\eta} = \frac{Q}{\pi K'} \frac{1}{\eta(\eta + 1)} \tag{236}$$

$$\frac{dx}{d\eta} = -\frac{Q}{\pi^2 K'} \frac{\ln \eta - (1/a)(\eta + 1)}{\eta(\eta + 1)} \tag{237}$$

Integrating these equations with $y(G) = 0$ yields

$$\frac{\pi K' y}{Q} = \ln \frac{\eta}{\eta + 1} \tag{238}$$

$$\frac{\pi K' x}{Q} = \ln \eta \left[\frac{1}{a} + \ln \frac{\eta + 1}{\eta^{1/2}} \right] + F(-\eta) + C$$

where $F(-\eta)$ is the dilogarithm function of η, or

$$F(\eta) = \int_o^1 \frac{\ln u}{u - 1/\eta}\, du$$

and C is determined by choosing the position of the origin. For the origin at A, $C = 0$.

The distance $\overline{AG}$ corresponds to

$$\overline{AG} = \frac{Q}{\pi^2 K'} \left[\int_1^\infty \frac{\ln \mu}{\mu(\mu - 1)}\, d\mu + \int_1^\infty \frac{d\mu}{a\mu} \right] = \frac{Q}{6K'} + \frac{Q \ln \mu}{\pi^2 K' a} \bigg|_1^\infty = \infty$$

The physical interpretation of this result is interesting. It means that this is the limit of a drain that intercepts only a portion of the total inflow Q while the remaining portion continues to flow as a thin wedge along the upper boundary of the flow region.

For any other value of $\beta < \pi/2$, we chose a point (S in Fig. 41a) through which the interface passes. This point determines the value of a by Eqs. (238) and (239), and, hence, β by Eq. (233).

Example 4. Partial interception by a sink above an interface. This example [*59*] is introduced to demonstrate a solution in which the hodograph plane takes the form of a Riemann surface. This case is similar to that considered in Example 3, except that the drain intercepts only part (Q_1) of the fresh water discharge, whereas the remaining part $Q_2 = Q - Q_1$ flows to the sea bottom. The physical z plane is shown on Fig. 42a. The interface (DM_2M_1F in Fig. 42a) has two inflection points: M_1 and M_2.

Figure 42b shows the hodograph $\tilde{\omega}$ plane of this flow. In this plane, the fresh water flow domain is mapped onto the region $BA_1A_2SG_1G_2FM_1M_2DCB$, which has the form of a Riemann surface [*62*]. The boundaries A_1A_2 and G_1G_2 are circles of infinite radii. The segment FM_1M_2 is a slot in the lower (shaded) sheet; there is a narrow "bridge" between the points SM_1 which connects the region FG_2G_1 to A_2. From A_2, over the large circle, to A_1, and toward B, an overlap encircles the slot at M_2. This overlap terminates at the beak that contains B and D. Thus, the flow domain in the $\tilde{\omega}$ plane is a double-sheet Riemann surface and cannot be cut out of a single sheet of paper. However, by gluing two sheets together, it is possible to make a model that is

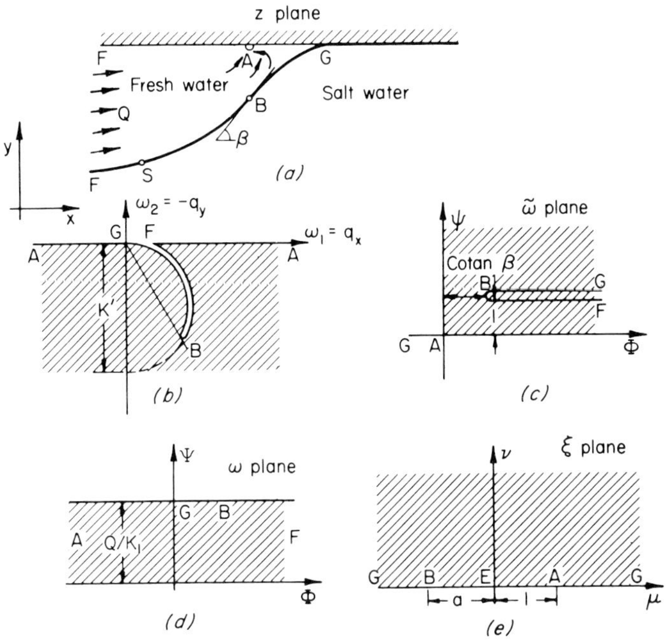

FIG. 41. Interface below a point sink in an infinitely thick aquifer [58].

simply connected. It is therefore possible to map this region on the upper half of the auxiliary ξ plane (Fig. 42*e*).

Figure 42*d* shows the complex potential ζ plane. The discharge rates Q_1 and Q_2 through the drain and through the outflow face *FG*, respectively, are considered known. The location of the stagnation point *S*, as yet unknown, is to be determined in the solution process.

To determine the shape of the interface in the *z* plane, we map the flow domain in both the ζ and the $\tilde{\omega}$ planes onto the upper half of the ξ (Fig. 42*g*). Then, Eq. (199) is applicable.

To map ζ on ξ we use the Schwarz–Christoffel transformation which leads to [59]

$$\zeta(\xi) = \alpha_2[\ln \eta - (a - s)a^{-1/2}(a + f)^{-1/2}L(\xi, f, a) + \beta_2 \tag{239}$$

where

$$\eta = [\xi^{1/2} + (t + f)^{1/2}]^2/f$$
$$L(\xi, f, a) = 2 \ln [\xi^{1/2}(a + f)^{1/2} + (\xi + f)^{1/2}a^{1/2}] - \ln f(a - \xi)$$

and the constants α_2, β_2, *a*, *f*, *s* have to be determined from the boundary conditions. Details are given in Ref. [59].

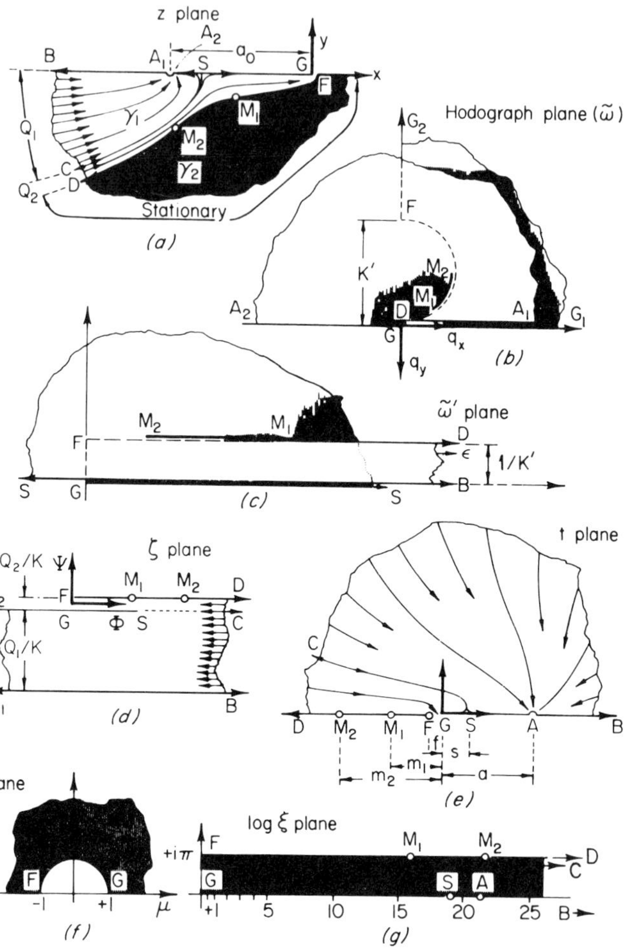

FIG. 42. Hodograph solution for a sink operating above an interface [59].

The $\tilde{\omega}'$ plane is obtained from the $\tilde{\omega}$ plane by the inverse transformation $\tilde{\omega}' = 1/\tilde{\omega}$. Then $\tilde{\omega}'$ is mapped onto the upper half of the ξ plane by

$$\tilde{\omega}' = \frac{1}{\pi K'}\left[\ln \xi - \frac{(a-s)t^{1/2}(t+f)^{1/2}}{(t-s)a^{1/2}(a+f)^{1/2}} \ln \xi_a\right] \tag{240}$$

where $\ln \xi_a = (m_1 + s)(m_2 + s)a^{1/2}(a+f)^{1/2}/s(s+f)(a-s)$, and the various constants are determined from the boundary conditions (for details see Ref. [59]).

Finally, these expressions are inserted in Eq. (199) from which the shape of the interface is derived in a parametric form.

Example 5. *Oil entrapment in an anticline*. This example [*64*] demonstrates a case in which the stationary liquid (γ_1) is the lighter one above the interface. Figure 43*a* shows the physical *z* plane. It is a case of a simplified anticlinal oil

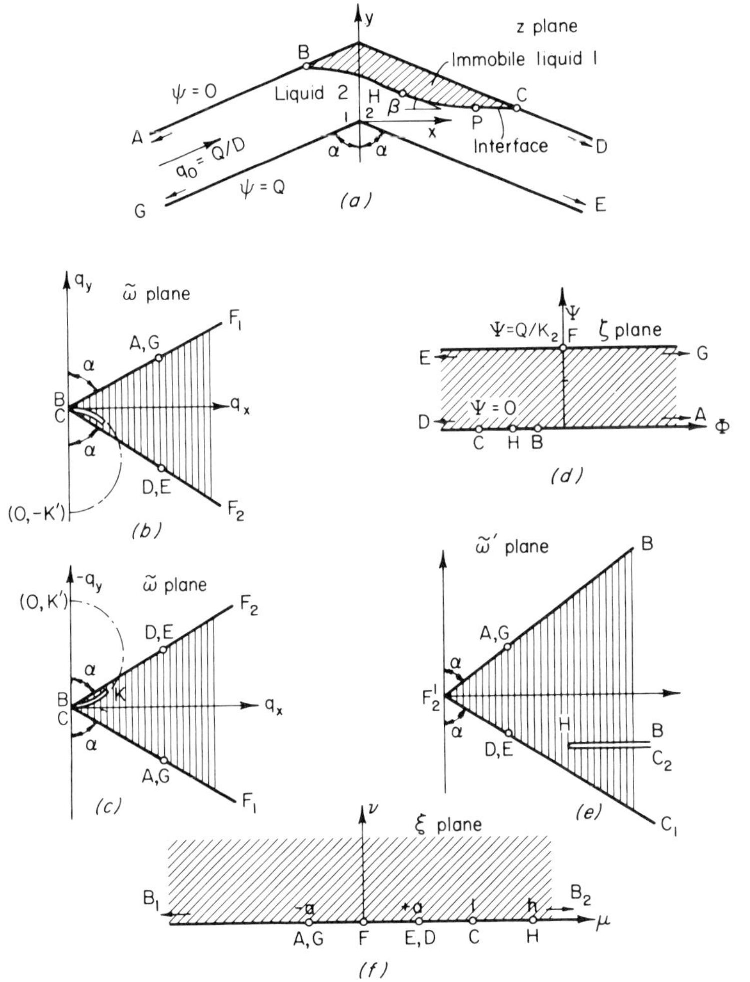

FIG. 43. An interface between a lower mobile fluid and an upper immobile one [*64*].

reservoir in which water is flowing under a stationary body of oil, assuming an abrupt interface between them. An inflection point usually appears at point *H*. Figures 43*b*, *c*, and *e* show the flow domain as mapped onto the ω, $\tilde{\omega}$, and $\tilde{\omega}' = K'/\tilde{\omega}$ planes where $K' = K_2(\gamma_2 - \gamma_1)/\gamma_2$. In the $\tilde{\omega}'$ plane, the

Across this zone the density of the mixed water varies from that of fresh water to that of sea water. However, under certain conditions the width of this zone is relatively small (e.g., when compared with the thickness of the aquifer), so that the abrupt interface approximation introduced in Sec. IV, A, is applicable. For example, observations [*65*] along the coast of Israel indicate that this assumption of an abrupt interface is indeed justified. Cooper [*66*] describes a case in which the transition zone is very wide so that the interface approximation is no more valid. In what follows we shall assume that an abrupt interface separates the regions occupied by the two fluids.

Under natural undisturbed conditions in a coastal aquifer a state of equilibrium is maintained, with a stationary interface and a fresh water flow to the sea. At every point along this interface the slope is determined by Eq. (57). The continuous change of slope (Fig. 44) results from the fact that as the sea is approached, q_{1s} increases. By pumping in excess of replenishment, the water table (or the piezometric surface) in the vicinity of the coast is lowered to the extent that the piezometric head in the fresh water body becomes less than in the adjacent sea water wedge, and the interface starts to advance inland. This phenomena is called *sea-water intrusion* (or *encroachment*). As the interface advances, the transition zone widens; however, we shall assume that the abrupt interface approximation remains valid. When the advancing interface reaches inland pumping wells, the latter become contaminated. Extensive research is being carried out in many parts of the world to understand the mechanism of sea water intrusion and to control it in order to improve the yield of coastal aquifers. Some elements of these investigations are reviewed here.

Actually, most of the theoretical background and solution approaches have already been presented in Sec. IV (especially Secs. IV, B and D) and will not be repeated here. The difficulties inherent in an exact analytical solution should by now be familiar to the reader. Some approximate approaches often employed by ground water hydrologists (in addition to those already discussed in Sec. IV, D) will now be given.

A tool that is relatively simple, yet most useful, for solving complicated interface management problems in coastal aquifers is the vertical Hele–Shaw analog described in Ref. [*24*, chap. 12]. Unless otherwise specified, from now on it will be assumed that the flow is two dimensional in a vertical plane.

B. The Ghyben–Herzberg Approximation

Beginning with Badon–Ghyben [*67*] and Herzberg [*68*], investigations of the coastal interface aimed both at determining the relationship between its shape and position and at determining the various hydrological components of a ground water balance in the region near the coast. Any coastal aquifer management scheme must be based on these relationships. Figure 45 shows the idealized Ghyben–Herzberg model of an interface in a coastal phreatic aquifer. Essentially, Ghyben and Herzberg assume static equilibrium with

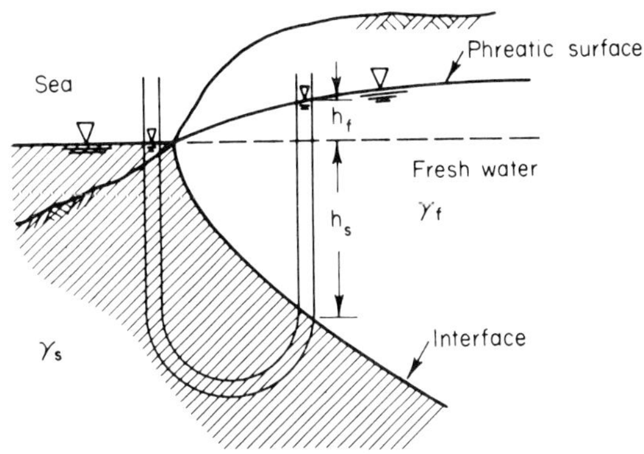

FIG. 45. Ghyben–Herzberg interface model.

stationary sea water and with a hydrostatic pressure distribution in the fresh water region. This means that equipotentials are vertical lines (or surfaces) and that the flow is horizontal (identical to the Dupuit assumptions). With the nomenclature of Fig. 45, we have under these conditions

$$h_s = \gamma_f/(\gamma_s - \gamma_f)h_f \equiv \delta h_f; \quad \delta = \gamma_f/(\gamma_s - \gamma_f) \tag{246}$$

For example, for $\gamma_s = 1.025$ gr/cm^3, $\gamma_f = 1.000$ gr/cm^3, $\delta = 40$, and $h_s = 40h_f$, i.e., at any distance from the sea, the depth of the interface below sea level is forty times the height of the fresh water table above it. Obviously, as the sea is approached the assumption of quasistatic equilibrium with horizontal flow is no longer valid, as the vertical flow component can be neglected no longer. Moreover, in Fig. 45 no outlet is left for the fresh water flow to the sea. Figures 45 and 46 show the actual steady flow conditions near the coast. Point

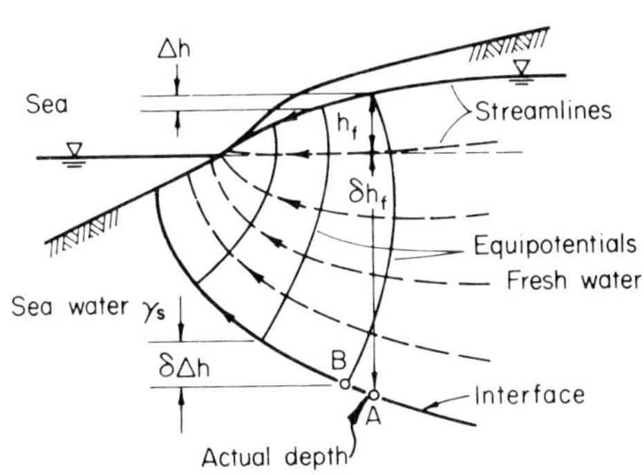

FIG. 46. Actual flow pattern near the coast.

A on the interface indicates the actual depth of the interface at that distance from the coast. Point B corresponds to the Ghyben–Herzberg approximation. Denoting h_s by y, we have, from Eq. (246),

$$y = -\delta h_f$$

If, in Eq. (63) we set $\varphi_{22} = 0$ (immobile sea water), we see that the difference between the Ghyben–Herzberg approximation (246) and the exact expression (61) stems from the difference between h_f and φ_{11} (i.e., between the assumed vertical equipotential with $\varphi_{11} = h_f$ and the actual curved one). In a confined aquifer, h_s in Eq. (246) is the depth below sea level of a point on the interface, whereas h_f is the (fresh water) piezometric head along a vertical line above this point.

Bear and Dagan [*39*, Rept. 1] investigated the validity of the Ghyben–Herzberg relationship. They found that in a confined horizontal aquifer of constant thickness D (Fig. 37*a*), the approximation is good (within an error of 5%) for determining the depth of the interface toe (point G in Fig. 37*a*), and hence the length of the intruding sea water wedge, provided $\pi K'D/Q > 8$ where Q is the fresh water discharge to the sea. Figure 47 shows this re-

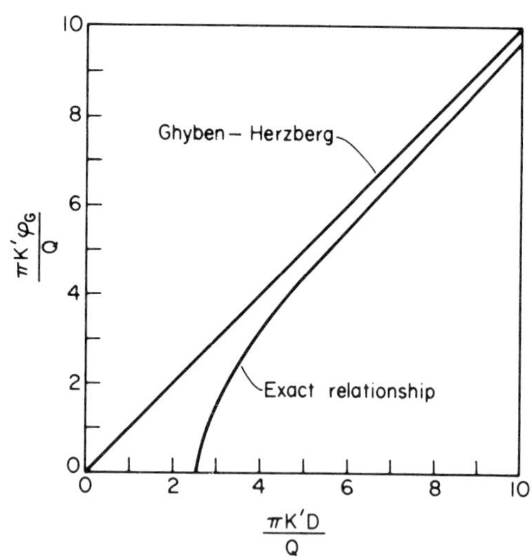

FIG. 47. Piezometric head above the interface toe in a confined aquifer.

lationship; the exact solution is obtained by the method of the hodograph (Sec. V, E, Example 1). The piezometric head φ_G is given by Eq. (217).

As the coast is approached, the depth of the interface is less than that given by the Ghyden–Herzberg relationship (Fig. 46). In the case of a phreatic aquifer, we must also have a seepage face above sea level. Note also that for a downward sloping flat sea bottom it follows from the discussion in Sec. V, C, 7 that the interface terminates at the sea bottom with a tangent in the vertical direction.

The (exact) conditions of dynamic equilibrium have already been dis-cussed in Sec. IV, A, 2. From Eq. (63) it follows that when the interface is in motion, we should also take into account variations in φ_{22} when determining the shape of the interface.

C. Determining the Shape of a Stationary Interface by the Dupuit–Ghyben–Herzberg Approximation [*69*]

Glover [*63*] used the similarity between Eqs. (57) and (58) and between the boundary conditions on a phreatic surface and on a stationary interface to determine the shape of the latter in an infinitely thick confined aquifer, using the method of complex variables (Fig. 48). The results were already presented

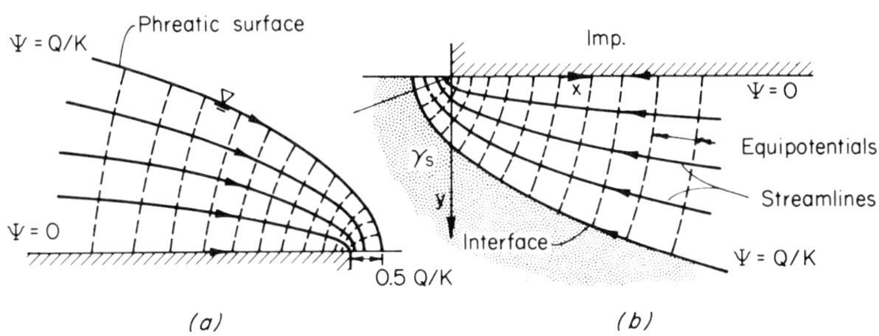

FIG. 48. The flow net near the coast.

in Sec. V, E, Ex. 1, Eqs. (214)—(216).

An exact solution for the shape of the interface can be derived for relatively simple geometries (confined aquifer, etc.) by the hodograph method dis-cussed in Sec. V. Example 1 of Sec. V, E, gives an exact solution for the case of a confined aquifer of constant thickness. Other examples of hodograph solutions for a coastal aquifer are given in Refs. [*57*, *58*, *59*]. However, for most practical cases (e.g., when the aquifer is a phreatic one, etc.)this approach is too complicated to be considered as a tool for engineering purposes. Such a tool is the combination of the Ghyben–Herzberg relationship with the Dupuit assumptions of horizontal flow.

1. *The Interface in a Confined Aquifer*

Consider the stationary interface in the horizontal confined aquifer shown in Fig. 49*a*. Let the $x = 0$ be located at the interface toe (point G). The sea-

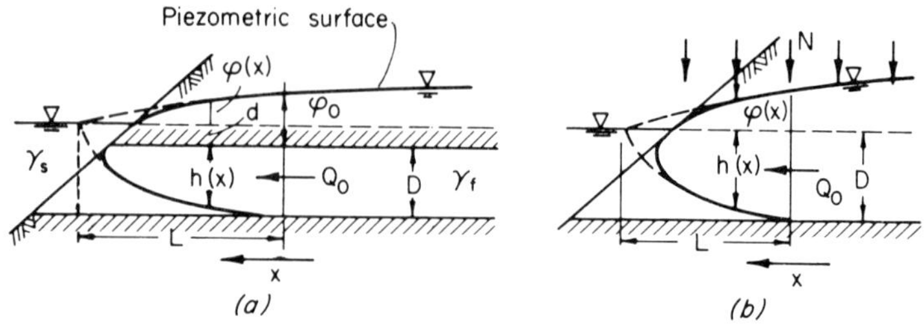

FIG. 49. The shape of the interface by the Dupuit–Ghyben–Herzberg approximation. (*a*) confined aquifer, (*b*) phreatic aquifer.

ward fresh water flow at this point is Q_o. Using Dupuit's assumption of horizontal flow of fresh water (and vertical equipotentials) continuity leads to

$$Q_o = \text{const.} = -Kh(x)\frac{\partial\varphi(x)}{\partial x} \tag{247}$$

where $K_f = k\gamma_f/\mu_f \equiv K$, $\varphi_f \equiv \varphi$. Since, by Eq. (246),

$$h_s = \delta\varphi_f = d + h(x), \qquad \delta\varphi_o = d + D$$

Eq. (247) becomes

$$Q_o = -\frac{kh}{\delta}\frac{dh(x)}{dx} \quad \text{or} \quad Q_o = -K(\delta\varphi - d)\frac{d\varphi(x)}{dx} \tag{248}$$

By integrating, with $x = 0$, $\varphi = \varphi_o$, $h = D$, we obtain

$$Q_o x = \frac{K}{2\delta}[D^2 - h^2(x)], \qquad Q_o x = \frac{K\delta}{2}(\varphi_o{}^2 - \varphi^2) - Kd(\varphi_o - \varphi) \tag{249}$$

which shows that the interface has the form of a parabola. At $x = L$, we set $\varphi = 0$, $\varphi_o = (D + d)/\delta$, and obtain

$$Q_o L = \frac{K(D^2 - d^2)}{2\delta} = \frac{K\varphi_0(\delta\varphi_0 - 2d)}{2} \tag{250}$$

2. *The Interface in a Phreatic Aquifer with Accretion*

Figure 49*b* shows a phreatic aquifer with accretion (N). Again, assuming that the flow in the aquifer is essentially horizontal and $h(x) = \delta\varphi(x)$, continuity leads to

$$Q_o + Nx = -K(h + \varphi)\frac{\partial\varphi}{\partial x} = -K(1 + \delta)\varphi\frac{\partial\varphi}{\partial x} \tag{251}$$

Integrating from $x = 0$, $\varphi = \varphi_0$, $h = D$, leads to

$$\varphi_o^2 - \varphi^2 = \frac{2Q_o x + Nx^2}{2K(1 + \delta)} \qquad (252)$$

At $x = L$, $\varphi = 0$, hence,

$$\varphi_o^2 = \frac{2Q_o L + NL^2}{K(1 + \delta)} \qquad (253)$$

For $N = 0$,

$$\varphi_o^2 = \frac{2Q_o L}{K(1 + \delta)} \qquad (254)$$

Again, the interface has a parabolic shape. In Figs. 49*a* and *b*, the dashed lines represent the Dupuit approximation.

From the practical point of view the most interesting outcome of Eqs. (249)–(254), is the relationship between Q_o and L. As Q_o increases, L decreases, and vice versa. This indicates that the length of sea water intrusion is a decision variable. By controlling Q_o we can control the length of desired sea water intrusion. The decision on the length of sea-water intrusion, or, equivalently, how much fresh water should be drained to the sea, is part of the coastal aquifer management problem. Equations (249)–(254) provide the tools that aid in making these decisions.

From Eqs. (250) and (253) we have a relationship between φ_o (piezometric head above toe), Q_o, and L. By controlling φ_o, say by means of artificial recharge, the water table may be lowered both landward and seaward of the toe without causing any additional sea intrusion. Landward of the toe, water levels may be changed according to some optimal management scheme. If pumpage is taking place seaward of the toe, the interface there will rise and may contaminate the wells if they are not at a sufficient distance above it. This conclusion underlies the idea of a *coastal collector*, which is a project executed by Water Planning for Israel Ltd. along the Mediterranean coast of Israel. The objective of the project is to increase the yield of the coastal aquifer by pumping through an array of shallow wells parallel to the coast and at a relatively small distance from it. For additional information the reader is referred to the literature on the project [*70*].

D. An Approximate Solution for the Moving Interface

Up to this point only a steady coastal interface has been considered. However, any change in the fresh water flow regimen is followed by a transition period during which the flow in the aquifer is unsteady and the interface is moving slowly toward its new equilibrium. The exact mathematical statement of the problem was presented in Sec. IV, A and B, and it was emphasized that to date an exact analytical solution is not possible. Some approximate approaches were presented in Secs. IV, D and F. Following a procedure similar to that leading to Eq. (129), Bear and Dagan [*39*, Rept. 2; *71*] presented approximate solutions for an interface moving in a coastal aquifer. Figure 50

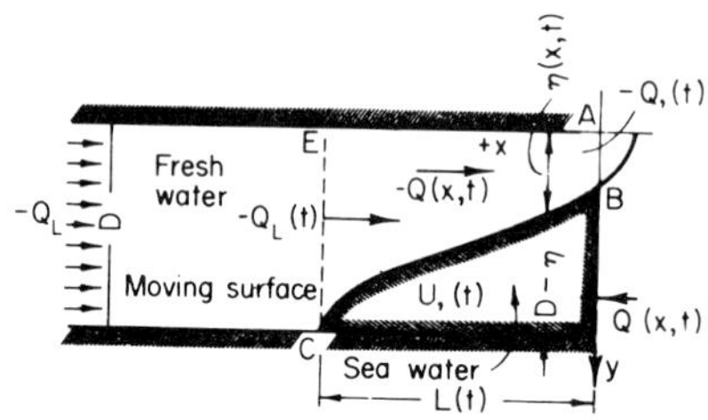

FIG. 50. Nomenclature for a moving interface.

gives the nomenclature for this solution. As the interface moves, the thickness η of the fresh water region varies with x and t. Initially, the seaward fresh water discharge $-Q_{Lo}$ is equal to the discharge to the sea. If at $t = t_o$, the discharge $-Q_{Lo}$ above the toe is suddenly reduced to a new constant $-Q_{Lf}(|Q_{Lf}| < |Q_{Lo}|)$, the interface moves landward until a new equilibrium is reached. For $Q_{Lf} \geq 0$, i.e., landward flow above the toe, no equilibrium can be reached. During the transition period, $Q_s(t)$ is reduced from $-Q_{so}$ to $-Q_{sf} = -Q_{Lf}$.

Assuming that the aquifer is thin, i.e., with L/D sufficiently large, the Dupuit assumptions may be used to express q_1, q_2, and $\sin\alpha$ which appear in Eq. (50) as

$$\sin\alpha = \partial\eta/\partial x, \qquad q_1 = Q(x,t)/\eta, \qquad q_2 = Q'(x,t)/(D-\eta) \tag{255}$$

where $Q'(x,t)$, $Q(x,t)$ (both positive) are the landward total salt water and fresh water discharges, respectively. Inserting Eq. (255) into Eq. (56) yields

$$\frac{\partial\eta}{\partial x} = \frac{Q'}{K_2'}\frac{1}{D-\eta} - \frac{Q}{K_1'}\frac{1}{\eta}, \qquad K_i' = \frac{k(\gamma_2-\gamma_1)}{\mu_i}, \quad i = 1(=f),\ 2\ (=s) \tag{256}$$

Or, since $Q + Q' = Q_L(t)$,

$$\frac{\partial\eta}{\partial x} = \frac{Q_L - Q}{K_2'(D-\eta)} - \frac{Q'}{K_1'\eta} \tag{257}$$

As the interface is displaced, Eqs. (121) and (122) are valid;

$$n\partial\eta/\partial t = -\partial Q/\partial x, \qquad n\partial\eta/\partial t = \partial Q'/\partial x \tag{258}$$

where n is effective porosity to displacement. Assuming for simplicity that $\mu_2 \approx \mu_1$, we have, by combining Eqs. (256) and (258),

$$n\frac{\partial\eta}{\partial t} = \frac{\partial}{\partial x}\left[\frac{K'\eta(D-\eta)}{D}\frac{\partial\eta}{\partial x} - \frac{Q_L\eta}{D}\right] \tag{259}$$

which is still nonlinear and identical to Eq. (129) with $b \rightarrow D$, $M = 1$, $Q_2 = bq$. Instead of attempting to solve Eq. (259), Bear and Dagan [*71*] followed the approximate quasisteady approach suggested by Polubarinova-Kochina [*53*]. In this approach, $Q(x, t)$ varies from Q_L at $x = L$, $\eta = D$, to Q_s at $x = 0$, $\eta \approx 0$, according to some predetermined rule. At the same time, Q' varies in a similar manner from zero to Q_L. The fresh water seaward discharge $Q_s(t) = Q(0, t)$ is regarded as the unknown variable of the problem. Continuity considerations applied to the volume of salt water $U_s(t)$ lead to

$$n(dU_s(t)/dt) = Q'(0, t) = Q_L - Q_s \tag{261}$$

where

$$U_s(t) = \int_{\eta=0}^{D} x(\eta, t)\, d\eta$$

Accordingly, the system of equations in the two unknowns Q_s and η are

$$\frac{\partial\eta}{\partial x} = \frac{Q_L}{K'(D-\eta)} - \frac{Q}{K'}\frac{D}{\eta(D-\eta)} \tag{261}$$

$$n\frac{d}{dt}\int_0^D x\, d\eta = Q_{Lf} - Q_s \tag{262}$$

In the proposed method, we assume some distribution of $Q = Q(Q_s, Q_L, \eta)$, solve Eq. (261) for $x = x(\eta, Q_s, Q_L)$, and then obtain $Q_s = Q_s(t)$ from Eq. (262). The accuracy of the solution depends entirely on the accuracy of the assumed variations of Q.

Two approximations were considered by Bear and Dagan.

(i). A linear variation of Q with η,

$$Q(\eta, t) = QL_f - [Q_{Lf} - Q_s(t)][D - \eta(x, t)]/D \tag{263}$$

and

(ii).
$$Q(\eta, t) = (1 - a)(Q_{Lf} - Q_s)(\eta/D) + Q_s \tag{264}$$

where a is a constant, ($\approx$0.1 from model experiments). Figure 51 shows the two approximations and the corresponding shapes of the resulting interfaces.

The first approximation leads to

$$\begin{aligned} \partial\eta(x, t)/\partial x &= -Q_s(t)/K' \\ \eta^2(x, t) &= -2Q_s(t)x/K'; \quad x = 0, \eta = 0 \end{aligned} \tag{265}$$

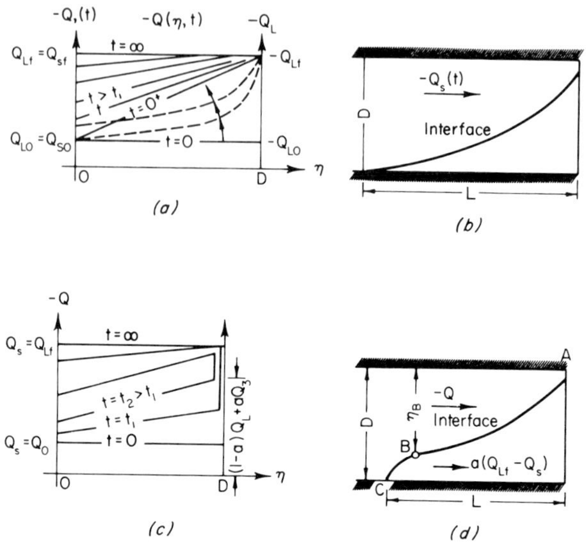

FIG. 51. First and second approximations for $Q = Q(\eta, t)$ and the resulting interface.

which is the well-known parabolic shape of the interface in steady flow. However, here a different parabola describes the interface at different times according to the variations in $Q_s(t)$. From Eq. (261) and $U_s = -K'D^3/6Q_s$, we obtain $Q_s(t)$.

From the second approximation,

$$\frac{\partial \eta}{\partial x} = \frac{a(Q_{Lf} - Q_s)}{K'(D - \eta)} - \frac{Q_s}{K'\eta} \tag{266}$$

which, by comparison with Eq. (256), may again be interpreted as describing at each instant the steady flow with constant Q_s. Here also, solutions $x = x(\eta, Q_s)$ and $Q_s = Q_s(t)$ may be obtained [*71*]. For $Q_{Lf} = 0$, i.e., stopping all seaward fresh water flow, it is possible to obtain the movement of the toe of the interface in the form

$$x\,|_{\eta=D} = (mK'D/n + x_o{}^2)^{1/2} \tag{267}$$

where x_o is the value of L at $t = 0$. For the first approximation, $m = 3$, for the second approximation $m = 1.75$. Rumer and Harleman [*72*] obtained $m = 1$.

Bear and Dagan [*39*, Rept. 2] carried out experimental work on a Hele–Shaw model [*24*, Chap. 12] to examine the validity of the approximate approaches. It seems that the first approximation is applicable with sufficient accuracy for practical purposes for a retreating interface only, while the second approximation describes rather well a landward advancing interface. In both cases, Q_s must remain sufficiently large. The results for $x(t)$ are sufficiently good for practical purposes in the range $\eta/D < 0.7$.

F. Interface Upconing

Under certain circumstances, pumping takes place from the region above an interface. This may happen as a result of sea water intrusion. It is also an intrinsic aspect of the coastal collector project just mentioned (Sec. VI, C). As a result of the pumping and the potential distribution established in the two fluids, the interface below the well (or a drain in two-dimensional flow) rises, at first slowly and then faster, toward the pumping well. Under certain conditions, a new equilibrium may be reached before the upconed interface reaches the pumping well. Otherwise, the upconed interface will eventually reach the pumping well in a cusplike form. When pumping stops, the upconed interface mound undergoes decay toward the initial steady state shape of the interface. When seaward flow of fresh water takes place, the upconed interface is displaced seaward as it rises and then decays. As was already explained, a transition zone rather than an abrupt interface separates the two miscible fluids, and this zone becomes wider as the interface moves. Nevertheless, the approximation of an abrupt interface will be used in the following paragraphs.

In (oil) reservoir engineering the same phenomenon, known as *water coning* [*2*, *8*, *30*], occurs when oil is withdrawn from a well located above the oil-water interface.

To date, only approximate (in addition to analog and numerical) solutions are known. Two such approximations will now be described. For the case of a stationary upconed interface, the hodograph method is applicable. An example is given in Sec. V, E, Ex. 2.

1. *The Muskat approximation* [*2*, *30*]

Muskat's approximation deals with steady flow and relatively small interface rises. Under these conditions Muskat assumes that the lower (heavier) liquid is stationary, and that the potential distribution in it is hydrostatic. He also assumes that the shape of local upconing of the interface beneath the sink (well or drain) does not appreciably affect the potential distribution in the region occupied by the lighter liquid. Strictly speaking, a stationary interface behaves like an impervious boundary whose form does affect the potential distribution in the flow domain. With these assumptions it is possible to derive the shape of the upconed interface from the potential distribution on some hypothetical horizontal plane introduced as an impervious boundary to the flow above it.

In Fig. 52, a point sink withdraws liquid 1 at a rate Q from a confined homogeneous and isotropic aquifer of thickness D. From Muskat's assumptions it follows that

$$\Delta\zeta = \zeta_M - \zeta_N = \frac{(\Phi_N - \Phi_M)\mu_1}{k(\gamma_2 - \gamma_1)} \cong \frac{(\Phi_{N'} - \Phi_{M'})\mu_1}{k(\gamma_2 - \gamma_1)} = \frac{\varphi_{N'} - \varphi_{M'}}{\Delta\gamma/\gamma_1} \quad (268)$$

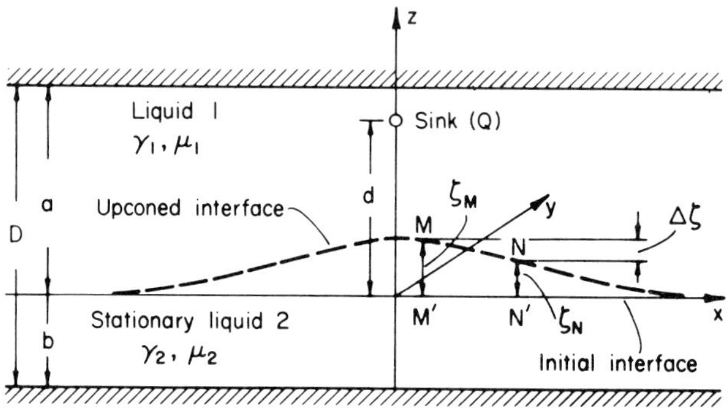

FIG. 52. Upconing by Muskat's approach.

where $\varphi_{M'}$ and $\varphi_{N'}$ are piezometric heads determined on the xy plane. This is also the assumption that underlies the linearized approximation based on small perturbations (Sec. IV, F, 1). In fact, the analytic considerations given there may serve as a justification for the Muskat approximation.

In Eq. (268) the steady state potential distribution $\varphi(x, y)$ on the hypothetical impervious boundary may be obtained analytically (for simple cases), numerically, or by means of an electric analog. Bear and Dagan [*39*, Rept.3] used an electric analog of the electrolytic tank type to determine $\varphi(x, y)$ and from it the shape of the upconed interface.

2. *Unsteady Upconing by the Method of Small Perturbations*

This approximation was already discussed in Sec. IV, F, 1. Bear and Dagan [*39*, Rept. 3] used this approach to determine the shape of the upconed interface in two- and three-dimensional flows to a sink pumping above an interface.

As an example, consider the case of a sink in the form of a drain located at a distance d above a horizontal interface. The flow is two-dimensional in the xz plane (Fig. 52). The solution is based on the nomenclature of Sec. IV, F, 1.

Mathematically stated, the problem is to solve the Laplace equations for φ_1 and φ_2 (or, since the medium is homogeneous and isotropic for Φ_1 and Φ_2), in regions 1 and 2. Since in this case φ_1°, φ_2°, and ζ° are identically zero, we have only to solve for the unsteady parts φ_1', φ_2', and ζ'. Henceforth for the sake of simplicity we shall denote $\varepsilon\varphi_1'$, $\varepsilon\varphi_2'$, and $\varepsilon\zeta'$ by φ_1, φ_2, and ζ. The heads φ_1 and φ_2 have to satisfy the following boundary conditions

$$\begin{aligned} z &= a, \qquad \partial\Phi_1/\partial z = 0 \\ z &= -b, \qquad \partial\Phi_2/\partial z = 0 \end{aligned} \qquad t \geq 0, \ -\infty < x < \infty \tag{269}$$

$$t \geq 0 \lim_{r\to 0} r(\partial\Phi_1/\partial r) = Q/2\pi; \quad r^2 = x^2 + (z-d)^2 \tag{270}$$

Along the interface, from Eq. (147),

$$\begin{aligned} \beta_1\, \partial\Phi_1/\partial t - \beta_2\, \partial\Phi_2/\partial t - \partial\Phi_1/\partial z &= 0 \\ \beta_1\, \partial\Phi_1/\partial t - \beta_2\, \partial\Phi_2/\partial t - \partial\Phi_2/\partial z &= 0 \end{aligned} \tag{271}$$

where $\beta_i = n\alpha_i$; $i = 1, 2$. The elevations of the interface are given by

$$\zeta = \alpha_2\Phi_2 - \alpha_1\Phi_1 \tag{272}$$

Hence, the initial condition is

$$t = 0, \qquad z = 0, \qquad \zeta = \alpha_2\Phi_2 - \alpha_1\Phi_1 = 0 \tag{273}$$

At infinity, the interface remains undisturbed. Hence, we have there

$$z = 0, \qquad x = \pm\infty, \qquad \zeta = \alpha_2\Phi_2 - \alpha_1\Phi_1 = 0 \tag{274}$$

Because of the singularity introduced by the drain, let the required solution for Φ_1 be decomposed into two parts, Φ_{11} and Φ_{12} (both satisfying the Laplace equation)

$$\Phi_1(x, z, t) = \Phi_{11}(x, z) + \Phi_{12}(x, z, t) \tag{275}$$

and such that Φ_{11} represents the singular character of Φ_1 at $x = 0$, $z = d$, whereas Φ_{12} is an unsteady state solution that is regular everywhere in region 1. Hence,

$$\Phi_{11} = \frac{Q}{4\pi} \ln [x^2 + (z-d)^2] - \frac{Q}{4\pi} \ln [x^2 + (z+d)^2] \tag{276}$$

The boundary conditions for Φ_{12} and Φ_2 are

$$z = a, \qquad \frac{\partial\Phi_{12}}{\partial z} = \frac{\partial\Phi_1}{\partial z} - \frac{\partial\Phi_{11}}{\partial z} = \frac{Q}{2\pi}\left[\frac{a-d}{x^2 + (a-d)^2} - \frac{a+d}{x^2 + (a+d)^2}\right] \tag{277}$$

$$z = -b, \qquad \partial\Phi_2/\partial z = 0 \tag{278}$$

$$\begin{aligned} z = 0, \qquad \Phi_{11} = 0, \qquad & \beta_1\, \partial\Phi_{12}/\partial t - \beta_2\, \partial\Phi_2/\partial t - \partial\Phi_{12}/\partial z = -Qd/\pi(x^2 + d^2) \\ & \beta_1\, \partial\Phi_{12}/\partial t - \beta_2\, \partial\Phi_2/\partial t - \partial\Phi_2/\partial z = 0 \end{aligned} \tag{279}$$

The initial conditions become

$$t = 0, \qquad z = 0, \qquad \zeta = \alpha_2 \Phi_2 - \alpha_1 \Phi_{12} = 0 \tag{280}$$

Let us now assume, since Φ_{12} and Φ_2 are everywhere regular, that they may be represented by Fourier integral transforms enabling separation of variables as

$$\Phi_{12} = \frac{1}{\pi} \int_0^\infty [A_1(\lambda, t) \exp(-\lambda z) + A_2(\lambda, t) \exp(\lambda z)] \cos(\lambda x)\, d\lambda \tag{281}$$

$$\Phi_2 = \frac{1}{\pi} \int_0^\infty A_3(\lambda, t) \cosh \lambda(z + b) \cos(\lambda x)\, d\lambda \tag{282}$$

In this form both Φ_{12} and Φ_2 satisfy the Laplace equation; A_1, A_2, and A_3 are functions to be determined by the boundary and the initial conditions (277), (279), and (280). The boundary condition (278) is satisfied by condition (282).

Using tables, the right-hand side of Eq. (277) and the first of Eqs. (279) are now represented in the form of a Fourier integral, similar to Eq. (281)

$$\frac{Q}{2\pi} \left[\frac{a - d}{x^2 + (a - d)^2} - \frac{a + d}{x^2 + (a + d)^2} \right] = \frac{Q}{2\pi} \int_0^\infty \{\exp[-\lambda(a - d)] - \\ - \exp[-\lambda(a + d)] \cos(\lambda x)\, d\lambda \tag{283}$$

$$\frac{Q}{2\pi} \frac{d}{x^2 + d^2} = \frac{Q}{\pi} \int_0^\infty \exp(-\lambda d) \cos(\lambda x)\, d\lambda \tag{284}$$

By inserting Eqs. (281)–(284) into Eqs. (277), (279), and (280), three equations in A_1, A_2, A_3, and λ are derived. Also for ζ,

$$\zeta = \frac{1}{\pi} \int_0^\infty [\alpha_2 A_3 \cos(\lambda b) - \beta_1(A_1 + A_2)] \cos(\lambda x)\, d\lambda \tag{285}$$

Denoting

$$\bar{\zeta} = \alpha_2 A_3 \cos(\lambda b) - \beta_1(A_1 + A_2) \tag{286}$$

we observe, in view of Eq. (285), that $\bar{\zeta}$ is the inverse transform of ζ, such that

$$\zeta = \frac{1}{\pi} \int_0^\infty \bar{\zeta}(\lambda, t) \cos(\lambda x)\, d\lambda \tag{287}$$

After some algebraic manipulations, we obtain from Eq. (286) a partial differential equation for $\bar{\zeta}$. By solving this equation and inserting the result in Eq. (287),

$$\zeta = \frac{\alpha_1 Q}{\pi} \int_0^\infty \frac{1}{\lambda} \frac{\cosh [\lambda(a-d)]}{\sinh (\lambda a)}$$

$$\times \left\{1 - \exp\left(\frac{-\lambda}{\beta_1 \coth (\lambda a) + \beta_2 \coth (\lambda b)}\right) t\right\} \cos (\lambda x)\, d\lambda \quad (288)$$

With $\mathscr{L} = \lambda a$, $\mathscr{Z} = \zeta/\alpha_1 Q$, $\mathscr{T} = t/\beta_1 a$, $\mathscr{B} = b/a$, $\mathscr{X} = x/a$, $\mathscr{D} = d/a$, and $M = \mu_2/\mu_1$, Eq. (288) becomes

$$\mathscr{Z} = \frac{1}{\pi} \int_0^\infty \frac{\cosh (1 - \mathscr{D})}{\mathscr{L} \sinh \mathscr{L}} \left\{1 - \exp\left[\frac{-\mathscr{L}\mathscr{T}}{(\coth \mathscr{L} + M \coth \mathscr{L}\mathscr{B})}\right]\right\} \cos (\mathscr{L}\mathscr{X})\, d\mathscr{L} \quad (289)$$

Because of symmetry, the crest of the upconing interface is at $\mathscr{X} = 0$. There, Eq. (289) becomes

$$\mathscr{Z}_{\text{crest}} = \frac{1}{\pi} \int_0^\infty \frac{\cosh \mathscr{Z} (1 - \mathscr{D})}{\mathscr{L} \sinh \mathscr{L}} \left\{1 - \exp\left[\frac{-\mathscr{L}\mathscr{T}}{(\coth \mathscr{L} + M \coth \mathscr{L}\mathscr{B})}\right]\right\} d\mathscr{L} \quad (290)$$

For a very thick aquifer, $a \to \infty$, $b \to \infty$, Eqs. (289) and (290) become

$$\mathscr{Z} = \frac{1}{2\pi} \ln \frac{(\mathscr{X}/\mathscr{D})^2 + (\mathscr{T}' + 1)^2}{(\mathscr{X}/\mathscr{D})^2 + 1}; \qquad \mathscr{T}' = \frac{t}{(\beta_1 + \beta_2) d} \quad (291)$$

$$\mathscr{Z}_{\text{crest}} = \frac{1}{\pi} \ln (\mathscr{T}' + 1) \quad (292)$$

This equation was also derived by Pilatovski [*35*].

Figure 53 shows ζ and ζ_{crest} for an infinitely thick aquifer in a graphical form. The results are presented for the case of an anisotropic aquifer ($k_x \neq k_z$) using the transformation given by Bear and Dagan in Ref. [*29*].

Bear and Dagan [*39*, Repts. 3 to 5] also apply this method to cases in which the initial interface makes an agle α with the $+x$ axis. They also performed sand-box model experiments from which they concluded that the range of validity of the approximate solution was up to displacements which at the crest of the upconing interface reach the value of one third the distance from the initial interface to the sink.

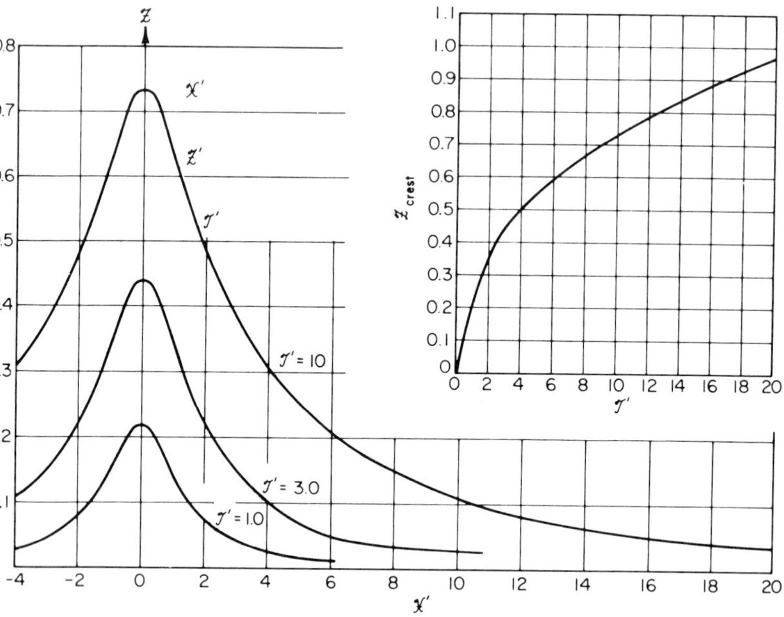

FIG. 53. The shape of the interface below a sink in two-dimensional flow.

$$\mathscr{X}' = \left(\frac{k_z}{k_x}\right)^{1/2}\frac{x}{d}, \quad \mathscr{Z} = \frac{(k_x k_z)^{1/2}(\gamma_2 - \gamma_1)}{\mu_1 Q}\,\zeta, \quad \mathscr{T} = \frac{k_z(\gamma_2 - \gamma_1)t}{n(\mu_1 + \mu_2)\,d}$$

Major Symbols

A	Area of cross section
b	Constant thickness of confined aquifer
D	Constant thickness of confined aquifer
$D(\ )/Dt$	$= d(\ )/dt$; material (or substantial) derivative
g	Gravity acceleration
$J(S_w)$	Leverett function
K	Hydraulic conductivity (dim L/T)
K_i	Hydraulic conductivity with respect to ith fluid
K_i'	$= K_i\,\Delta\gamma/\gamma_i$
k	Permeability (dim L^2)
$k_i, k^{(i)}$	Effective permeability to fluid i
$k_{ri}, k_r^{(i)}$	Relative permeability to fluid i (dimensionless)
L	Horizontal distance; length of sea water intrusion
M	Mobility ratio (ratio of (k/μ) of displacing fluid to (k/μ) of displaced fluid)
N	Accretion (dim L/T)
n	Porosity
p	Pressure
p_c	Capillary pressure
Q	Discharge
q	Specific discharge
r	Radial distance; radius
S	Saturation
S_e	Effective (or reduced) saturation

S_{nwo}	Residual nonwetting fluid saturation
S_{wo}	Irreducible wetting fluid saturation
s	Length along curve
t	Time
U	Volume
V	Velocity
x, y, z	Cartesian coordinates in three dimension (z vertically upward)
z	$= x + iy$; complex variable in xy plane
α_i	$= \mu_i/k\, \Delta\gamma$
α	Angle
β	Angle
γ	Specific weight
δ	$= \gamma_f/(\gamma_s - \gamma_f) = \gamma_f/\Delta\gamma$
ϵ	A small parameter
ζ	$= f(z)$; elevation of point on interface
η	Elevation of point on an interface in a vertical xy plane; thickness of flow domain above or below an interface
θ	Contact angle
μ	Dynamic viscosity
ν	Kinematic viscosity
ρ	Density; radial distance
σ_{ik}	Interfacial tension
σ_i	Surface tension
τ	Dimensionless time
φ	$= z + p/\gamma$; piezometric head
φ^*	$= z + \int_{p_0}^{p} dp/g\rho(p)$
Φ	$= K\varphi$; specific discharge potential
Ψ, ψ	Stream functions ($\Psi = K\psi$)
ω	$= q_x + iq_y$; complex specific discharge
$\bar{\omega}$	$= q_x - iq_y$; conjugate complex specific discharge

Subscripts

e	Effective; external
f	Fresh water
i	ith fluid; component in ith direction
n	Direction of normal
nw	Nonwetting fluid
o	Oil
r	Relative; in the r direction
s	In the s direction; salt water
w	Water; wetting fluid
x, y, z	Directions, components in these directions

References

1. Scheidegger, A. E., "The Physics of Flow Through Porous Media." 2nd ed., Univ. of Toronto Press, Toronto, 1960.
2. Muskat, M., "Physical Principles of Oil Production." McGraw-Hill, New York, 1949.
3. Marle, C., "Cours de Production, Tome IV, Les Écoulement Polyphasiques." Institut Français du Pétrole, 1965.
4. Oroveanu, T., "Flow of Multiphase Fluids Through Porous Media." Editura Academiei Republicii Socialiste Romania, Bucarest, 1966.
5. Prandtl, L., and Tietjens, O. G., "Fundamentals of Hydro- and Aeromechanics." Dover, New York, 1957.
6. Bear, J., and Bachmat, Y., Hydrodynamic dispersion in non-uniform flow through porous media, taking into account density and viscosity differences. Technion, Israel. Inst. of Techn., Hydraulic Lab., P.N. 4/66 (In Hebrew with English summary), 1966.
7. Bikerman, J. J., "Surface Chemistry—Theory and Application," 2nd ed., Academic Press, New York, 1958.
8. Pirson, S. J., "Oil Reservoir Engineering." McGraw-Hill, New York, 1958.
9. Leverett, M. C., Capillary behavior in porous media, *Trans. AIME* **142**, 151 (1941).
10. Brooks, R. H., and Corey, A. T., Hydraulic properties of porous media. Hydrology Papers, No. 3, Colorado State Univ., Fort Collins, Colorado, 1964.

11. Chatenever, A., and Calhoun, J. C. Jr., Visual examination of fluid behavior in porous media. *Trans. AIME* (*Petroleum Div.*) **195**, 149–156 (1952).
12. Hubbert, M. K., Darcy's law and the fluid equations of the flow of underground fluids. *Trans. Am. Inst. Min. Metal. Eng.* **207**, 222–239 (1956).
13. Wyckoff, R. D., and Botset, H. G., The flow of gas-liquid mixtures through unconsolidated sands. *Physics* **7**, 325–345 (1936).
14. Muskat, M., Wyckoff, R. D., Botset, H. G., and Meres, M. W., Flow of gas-liquid mixtures through sands. *Trans. AIME* (*Petroleum Div.*) **123**, 69–96 (1937).
15. Russell, T. W. F., and Charles, M. E., Effect of the less viscous liquid in the laminar flow of two immiscible liquids. *Can. J. Chem. Eng.* **37**, 18–24 (1959).
16. Yuster, S. T., Theoretical considerations of multiphase flow in idealized capillary systems, *Proc. 3rd World Petrol. Congr.* II, 437–445, The Hague, 1953.
17. Milne-Thomson, L. M., "Theoretical Hydrodynamics," 3rd ed. Macmillan, New York, 1955.
18. Hall, W. A., An analytical derivation of Darcy's equation. *Trans. Am. Geophys. Un.* **37**, 185–188 (1956).
19. Scott, P. H., and Rose, W., An explanation of the Yuster effect. *Petr. Tech.* **5** (1), 19–20 (1953).
20. Rose, W., Fluid flow in petroleum reservoirs. III. Effect of fluid-fluid interfacial boundary condition. *Ill. Geol. Surv. Circ.* **291**, 18 (1960).
21. Odeh, A. S., Effect of viscosity ratio on relative permeability. *Trans. AIME* **216**, 346–352, and discussion by C. F. Wienaug, 352–353 (1959).
22. Burdine, N. T., Relative permeability calculations from pore-size distribution data. *Trans. AIME* (*Petroleum Div.*) **198**, 71–78 (1953).
23. Stallman, R. W., Flow in the zone of aeration, *Advan. Hydrosci.* **4**, 151–195, 1967.
24. Bear, J., Zaslavsky, D., and Irmay, S., "Physical Principles of Water Percolation and Seepage." Unesco, Paris, 1968
25. Buckley, S. E., and Leverett, M. C., Mechanism of fluid displacement in sands. *Trans. AIME* **146**, 107 (1942).
26. Welge, H., A simplified method for computing oil recovery by gas or water drive. *Trans. AIME* **195**, 91 (1952).
27. Hubbert, K. M., The theory of ground water motion, *J. Geol.* **48**, 785–944 (1940).
28. Lusczynski, N. S., Head and flow of groundwater of variable density, *J. Geophys. Res.* **66** (12), 4247–4256 (1961).
29. Bear, J., and Dagan, G., The relationship between solutions of flow problems in isotropic and anisotropic soils. *J. Hydrology* **3**, 88–96 (1965).
30. Muskat, M., "The Flow of Homogeneous Fluids Through Porous Media." McGraw-Hill, New York, 1937.
31. Bear, J., and Jacobs, M., On the movement of water bodies injected into aquifers. *J. Hydrology* **3** (1), 37–57 (1965).
32. Le Fur, B., and Sourieau, Pl., Étude de l'écoulement diphasique dans une couche inclinée et dans un modèle rectangulaire de milieu poreux. *Rev. Inst. Franç. Pétrole* **18**, 325, 343 (1963).
33. Dietz, D. N., A theoretical approach to the problem of encroaching and by-passing edge water. *Proc. Kononkl. Ned. Akad. Wetenschap.* **B56** (1), 83–92 (1953).
34. Jacquard, P., Calculs numérique de déplacement de fronts. Rapport, Institute Français du Pétrole, ref. 7844, 1962.
35. Pilatovski, V. P., On the problem of ground water encroachment to a drain in a layer of infinite thickness (in Russian). *Isv. Akad. Nauk SSSR., Otd. Tech. Nauk.* **7**, 70–75 (1958).

36. Pilatovski, V. P., On the displacement of the interface between two heavy fluids in a horizontal layer (in Russian). *Ak. Nauk. SSSR, O.T.N. Mech. i Maschin* **1**, 127–130 (1961).
37. De Josselin de Jong, G., Singularity distribution for the analysis of multiple fluid flow through porous media. *J. Geophys. Res.* **65**, 3739–3758 (1960).
38. Aronofsy, J. S., and Ramey, H. J., Mobility ratio—its influence on injection or production histories in five spot water flood. *Trans. AIME* **207**, 205 (1956).
39. Bear, J., and Dagan, G., The transition zone between fresh and salt waters in a coastal aquifer. Prog. Rep. 1: The Steady interface between two immiscible fluids in a two-dimensional field of flow, 1962; Prog. Rep. 2: A steady flow to an array of wells above the interface—Approximate solution for a moving interface, 1963; Prog. Rep. 3: The unsteady interface below a coastal collector, 1964; Prog. Rep. 4: Increasing the yield of a coastal collector by means of special operation techniques, 1966; Prog. Rep. 5: The transition zone at the rising interface below the collector, 1966; Hydraulic Lab., Technion-Israel Institute of Technology, Haifa, Israel.
40. Dagan, G., Second order linearized theory of free surface flow in porous media. *La Huille Blanche* **8**, 901–910 (1964).
41. Stoker, J. J., "Water Waves." Wiley (Interscience), New York, 1957.
42. Kidder, R. E., Motion of the interface between two immiscible liquids of unequal density in a porous medium. *J. Appl. Phys.* **27**, 1546–1548 (1956).
43. Cailleau, J., Chaumet, P., Jeanson, B., and Van Quy, N., Étude numérique et expérimentale de l'edge coning. *Rev. Inst. Franç Pétrole* **18**, 344–357 (1963).
44. Dagan, G., The movement of the interface between two liquids in a porous medium with applications to a coastal aquifer. Ph.D. Thesis (in Hebrew). Technion, Haifa, Israel, 1964.
45. Scheidegger, A. E., Statistical hydrodynamics in porous media. *Advan. Hydrosci.* **1**, 161–181, 1964.
46. Chouke, R. L., van Meurs, P., and van der Poel, C., The instability of slow, immiscible, viscous liquid-liquid displacements in permeable media. *J. Petrol. Techn.* **11**, 64, 73 (1959).
47. Jacquard, P., and Seguier, P., Mouvement de deux fluides en contact dans un milieu poreux. *J. Mech.* **1** (*4*), 367–394 (1962).
48. Chouke, R. L., van Meurs, P., and Van der Poel, C., The instability of slow, immiscible, viscous liquid-liquid displacement in permeable media. *Trans. AIME* **216** (*Petroleum Div.*), 188–194 (1959).
49. Hamel, G., Uber Grundwasserstromung. *Z. Angew. Math. Mech.* **14**, 129 (1934).
50. Dachler, R., "Grundwasserstromung." Springer, Wien, 1936.
51. Davison, B. B., The flow of ground water through an earth-filled cofferdam with vertical walls (in Russian). *Zapiski Gosudanstvennogo Gidrologicheskogo Inst.* **6**, 121 (1932).
52. Vedernikov, V. V., "Seepage from Channels" (in Russian). Gosstoÿzdat, Moscow, 1934.
53. Polubarinova-Kochina, P. Ya., "Theory of Ground Water Movement" (in Russian). Gostekhizdat, Moscow, 1952 (also English translation by R. J. M. DeWiest, Princeton Univ. Press, Princeton, New Jersey, 1962).
54. Aravin, V. I., and Numerov, S. N., "Theory of Motion of Liquids and Gases in Undeformable Porous Media" (in Russian). Gostekhizdat, Moscow, 1953 (also English translation by A. Moscona, Israel Program for Scientific Translation, Jerusalem, 1965).
55. Harr, M. E., "Groundwater and Seepage." McGraw-Hill, New York, 1962.
56. Bear, J., and Dagan, G., The use of the hodograph method for ground water investigations. Technion, Haifa, Dept. of Civil Eng., 1962.

57. Henry, H. R., Salt water intrusion into fresh water aquifers, *J. Geophys. Res.* **64** (*11*), 1911–1919 (1959).
58. Bear, J., and Dagan, G., Some exact solutions of interface problems by means of the hodograph method, *J. Geophys. Res.* **69** (21), 1563–1572 (1964).
59. De Josselin de Jong, G., A many valued hodograph in an interface problem. Technische Hogeschool Delft, Atd., Weg-en Waterbouwkunde, 1964.
60. Kidder, R. E., Flow of immiscible fluids in porous media: an exact solution of a free boundary problem. *J. Appl. Phys.* **27** (8), (1956).
61. Van Quy, N., Étude numérique et experiméntale de l'edge coning. *Rev. Inst. Franç. Pétrole* **18** (1963).
62. Churchill, R. V., "Introduction to Complex Variable and Applications." McGraw-Hill, New York, 1948.
63. Glover, R. E., The pattern of fresh water flow in a coastal aquifer. *J. Geophys. Res.* **64**, 439–475 (1959).
64. Braester, C., On the possibility of oil and gas entrapment in the north Negev region. M.Sc. Thesis (in Hebrew), Technion, Haifa, 1966.
65. Jacobs, M., and Schomorak, S., Sea water intrusion and interface determination along the coastal plain of Israel. State of Israel, Hydrological Service. Hydrological Paper No. 6, 1960.
66. Cooper, H. H., A hypothesis concerning the dynamic balance of fresh water and salt water in a coastal aquifer. *J. Geophys. Res.* **64**, 461–467 (1959).
67. Badon-Ghyben, W., "Nota in Verband met de Voorgenomen Butboring Nabij Amsterdam." (in Dutch), The Hague, Tydschrift van het Koninklyk Inst. van Ing., 21, 1888.
68. Herzberg, A., Die Wasserversorgung einiger Nordseebader (the water supply in parts of the North Sea Coast, in Dutch). *J. Gasbeleucht. Wasserversorg.*, **44**, 815–819, 842–844 (1901).
69. Santing, G., Investigations of ground water in the coastal plain of Israel. Water Planning for Israel Ltd., P.N. 23, 1957.
70. Joint experimental coastal ground water collector project. Final Rept., Water Planning for Israel Ltd., Tel Aviv, Israel, P.N. 445, 1965.
71. Bear, J., and Dagan, G., Moving interface in coastal aquifers. *Proc. ASCE* (*Hydrol. Div.*), **90** (4), 193–215 (1964).
72. Rumer, R. R., Jr., and Harleman, D. R. F., Intruded salt-water wedge in porous media. *Proc. ASCE* (*Hydrol. Div.*) **89** (6), 193–220 (1963).

AUTHOR INDEX

Numbers in parentheses are reference numbers and indicate that an author's work is referred to although his name is not cited in the text. Numbers in italics show the page on which the complete reference is listed.

SUBJECT INDEX